VOLUME SIXTY

ADVANCES IN
ECOLOGICAL RESEARCH

Resilience in Complex Socio-ecological Systems

ADVANCES IN ECOLOGICAL RESEARCH

Series Editors

DAVID A. BOHAN

Directeur de Recherche
UMR 1347 Agroécologie
AgroSup/UB/INRA
Pôle GESTAD, Dijon, France

ALEX J. DUMBRELL

School of Biological Sciences
University of Essex
Wivenhoe Park, Colchester
Essex, United Kingdom

VOLUME SIXTY

ADVANCES IN
ECOLOGICAL RESEARCH
Resilience in Complex Socio-ecological Systems

Edited by

DAVID A. BOHAN
Directeur de Recherche
UMR 1347 Agroécologie
AgroSup/UB/INRA
Pôle GESTAD, Dijon, France

ALEX J. DUMBRELL
School of Biological Sciences
University of Essex
Wivenhoe Park, Colchester, Essex,
United Kingdom

ACADEMIC PRESS
An imprint of Elsevier

ELSEVIER

Academic Press is an imprint of Elsevier
125 London Wall, London, EC2Y 5AS, United Kingdom
The Boulevard, Langford Lane, Kidlington, Oxford OX5 1GB, United Kingdom
525 B Street, Suite 1650, San Diego, CA 92101, United States
50 Hampshire Street, 5th Floor, Cambridge, MA 02139, United States

First edition 2019

Notices
Knowledge and best practice in this field are constantly changing. As new research and
experience broaden our understanding, changes in research methods, professional practices,
or medical treatment may become necessary.

Practitioners and researchers must always rely on their own experience and knowledge in
evaluating and using any information, methods, compounds, or experiments described
herein. In using such information or methods they should be mindful of their own safety and
the safety of others, including parties for whom they have a professional responsibility.

To the fullest extent of the law, neither the Publisher nor the authors, contributors, or editors,
assume any liability for any injury and/or damage to persons or property as a matter of
products liability, negligence or otherwise, or from any use or operation of any methods,
products, instructions, or ideas contained in the material herein.

ISBN: 978-0-08-102854-4
ISSN: 0065-2504

For information on all Academic Press publications
visit our website at https://www.elsevier.com/books-and-journals

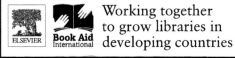

Working together
to grow libraries in
developing countries

www.elsevier.com • www.bookaid.org

Publisher: Zoe Kruze
Acquisition Editor: Jason Mitchell
Editorial Project Manager: Joanna Collett
Production Project Manager: Abdulla Sait
Cover Designer: Greg Harris

Typeset by SPi Global, India

Contents

Contributors

Craig R. Allen
U.S. Geological Survey, Nebraska Cooperative Fish and Wildlife Research Unit, School of Natural Resources, University of Nebraska—Lincoln, Lincoln, NE, United States

David G. Angeler
Department of Aquatic Sciences and Assessment, Swedish University of Agricultural Sciences, Uppsala, Sweden; School of Natural Resources, University of Nebraska—Lincoln, Lincoln, NE, United States

Noa Avriel-Avni
Dead Sea and Arava Science Center, Masada, Israel

Nesha Beharry-Borg
Centrascape Mid Centre Mall Compound, Chaguanas, Trinidad

Daniel S. Chapman
Biological and Environmental Sciences, Faculty of Natural Science, University of Stirling, Stirling, United Kingdom

Wen-Ching Chuang
National Risk Management Research Laboratory, U.S. Environmental Protection Agency, Cincinnati, OH, United States

Jan Dick
Centre for Ecology and Hydrology, Penicuik, Scotland, United Kingdom

Victoria M. Donovan
Department of Agronomy & Horticulture, University of Nebraska, Keim Hall, Lincoln, NE, United States

Tarsha Eason
National Risk Management Research Laboratory, U.S. Environmental Protection Agency, Research Triangle Park, NC, United States

Nico Eisenhauer
German Centre for Integrative Biodiversity Research (iDiv) Halle-Jena-Leipzig; Institute of Biology, Leipzig University, Leipzig, Germany

Benjamin Feit
Department of Ecology, Swedish University of Agricultural Sciences, Uppsala, Sweden

Evan D.G. Fraser
Department of Geography, Environment, and Geomatics, University of Guelph, Guelph, ON, Canada

Hanna Friberg
Department of Forest Mycology and Plant Pathology, Swedish University of Agricultural Sciences, Uppsala, Sweden

Hannah B. Fried-Petersen
Department of Aquatic Sciences and Assessment, Swedish University of Agricultural Sciences, Uppsala, Sweden

Ahjond Garmestani
National Risk Management Research Laboratory, U.S. Environmental Protection Agency, Cincinnati, OH, United States

Klaus Hubacek
Center for Energy and Environmental Sciences (IVEM), Energy and Sustainability Research Institute Groningen (ESRIG), University of Groningen, Groningen, The Netherlands; International Institute for Applied Systems Analysis, Laxenburg, Austria

Nanlin Jin
Department of Computer and Information Sciences, Northumbria University, Newcastle upon Tyne, United Kingdom

Mattias Jonsson
Department of Ecology, Swedish University of Agricultural Sciences, Uppsala, Sweden

Emily A. Martin
Department of Animal Ecology and Tropical Biology, Biocenter, University of Würzburg, Würzburg, Germany

Claire H. Quinn
Sustainability Research Institute, School of Earth and Environment, University of Leeds, Leeds, United Kingdom

Thomas Reitz
German Centre for Integrative Biodiversity Research (iDiv) Halle-Jena-Leipzig, Leipzig; Department of Soil Ecology, Helmholtz Centre for Environmental Research—UFZ, Halle, Germany

Fabrice Requier
Department of Animal Ecology and Tropical Biology, Biocenter, University of Würzburg, Würzburg, Germany

Caleb P. Roberts
Department of Agronomy & Horticulture, University of Nebraska, Keim Hall, Lincoln, NE, United States

Martin Schädler
German Centre for Integrative Biodiversity Research (iDiv) Halle-Jena-Leipzig, Leipzig; Department of Community Ecology, Helmholtz-Centre for Environmental Research—UFZ, Halle, Germany

Elke Schulz
Department of Soil Ecology, Helmholtz Centre for Environmental Research—UFZ, Halle, Germany

Julia Siebert
German Centre for Integrative Biodiversity Research (iDiv) Halle-Jena-Leipzig; Institute of Biology, Leipzig University, Leipzig, Germany

Shana M. Sundstrom
School of Natural Resources, University of Nebraska—Lincoln, Lincoln, NE, United States

Mette Termansen
Food and Resource Economics, University of Copenhagen, Frederiksberg C, Denmark

Madhav P. Thakur
German Centre for Integrative Biodiversity Research (iDiv) Halle-Jena-Leipzig; Institute of Biology, Leipzig University, Leipzig, Germany; Department of Terrestrial Ecology, Netherlands Institute of Ecology (NIOO-KNAW), Wageningen, The Netherlands

Dirac Twidwell
Department of Agronomy & Horticulture, University of Nebraska, Keim Hall, Lincoln, NE, United States

Alexandra Weigelt
German Centre for Integrative Biodiversity Research (iDiv) Halle-Jena-Leipzig; Department of Systematic Botany and Functional Biodiversity, Institute of Biology, Leipzig University, Leipzig, Germany

Carissa L. Wonkka
Department of Agronomy & Horticulture, University of Nebraska, Keim Hall, Lincoln, NE, United States

Rui Yin
Institute of Biology, Leipzig University, Leipzig; Department of Community Ecology, Helmholtz-Centre for Environmental Research—UFZ, Halle, Germany

Preface

A central question for Ecology, now and in the future, is how do we render or assure that the ecosystems that humans rely upon are resilient to environmental change? Ongoing global change, wrought by modifications, for example, to climate and land use, has placed considerable stress on the planet's ecosystems and underscored the need to measure ecosystem resilience, to understand mechanisms behind resilience and to provide appropriate management to assure that our ecosystems continue to deliver the services we require. This volume was conceived with certain constraints. Resilience was not to be limited to the ecology of ecosystems. Rather, resilience should also consider aspects of social resilience, including those of the differing sentiment and need between groups of stakeholders and the economics of resilience. There was no attempt, however, to limit the volume to a single definition of resilience such as ecological or engineering resilience (Holling, 1973, 1996), but to examine something of the range of work that is ongoing in this domain. The work covered here includes theoretical work outlining how resilience and correlates of stability might be measured theoretically, to pragmatic approaches working with various stakeholders, and improvements to measures of systems performance from those that currently exist.

In this volume entitled 'Resilience in Complex Socio-ecological Systems', we present five articles that illustrate something of the diversity of work currently being undertaken from first-principles theoretical work on the definition and measurement of socioecological resilience, via experimental systems to practical case studies in real-world systems. In Chapter 1, David Angeler and colleagues argue for measuring 'adaptive capacity' as an approach to resilience. In their chapter, the authors detail past definitions of adaptive capacity, which have often been confusing, to define the concept as 'the latent potential of an ecosystem to alter resilience in response to change'. Their now operational concept is then used to propose hypotheses that through modelling and appropriate sampling could be tested and evaluated. Angeler et al. believe that this would, over time, incrementally reduce key uncertainties about system resilience and improve our understanding of the adaptive capacity of ecosystems.

Julia Siebert and her coauthors then examine the resilience of soil-derived ecosystem functions in grasslands, including nutrient cycling and decomposition, from an experimental system designed to examine the potential effects of climate change. Siebert et al. (Chapter 2) explicitly examine both the ecological insurance hypothesis, that a higher diversity of soil communities would assure more stable provisioning of ecosystem functions under climate change, and past support for this idea in grasslands. Their 2-year study shows that the manipulation of temperature and humidity conditions, which mimic future climate, consistently reduced soil biological activity. Manipulations that increased land-use intensity also had reduced soil biological activity, but in contrast to past results, extensification of land use did not alleviate the detrimental effects of simulated climate change. Rather, the extensively used grasslands showed the greatest relative reduction in soil biological activity, highlighting vulnerability and the need to identify complementary managements to assure their future functioning.

In Chapter 3 Martin et al. explore whether it is possible to operationalize resilience to biodiversity-derived functions in agroecosystems under environmental change. This question is of great importance for the long-term security and environmental safety of these systems, but because of the marked conceptual differences that exist between different measures, resilience operationalization is difficult. These authors use a four-step approach of summarizing potential dimensions of resilience and disturbance under environmental change, reviewing indicators of the resilience of biodiversity-driven functions in agroecosystems, using these indicators to explore the resilience of the ecosystem services of biological pest control, biological disease control in soil and pollination, and finally reviewing those tools and approaches that could catalyse further the assessment of resilience of biodiversity-driven agroecosystem functions.

Mette Termansen then explores how understanding the coupling of social and ecological systems is a requisite to attain sustainable and resilient land management in the long term. Termansen et al. (Chapter 4) contrast a standard approach to understanding land-use change, which is based upon those land-manager behaviours applied when managed areas are influenced by direct and indirect drivers of environmental change, the subsequent resulting changes in land use, and the environmental consequences of that change, with an alternative approach. This proposes the combined use of land-use dynamics modelling, integrating land management behavioural models derived from choice experiments, and spatially explicit systems dynamics modelling. The advantages of this alternative are that it explicitly models potential environmental and management feedbacks, that behavioural rules can be tested from actual and hypothetical choice data, and that the behavioural choice models generate probabilities of alternative behaviours for realistic scenarios of the future.

In Chapter 5 by Avriel-Avni and Dick, the authors present a case study of how scientists and land-managers can have differing views of same system. This question is important because acceptable, codeveloped management of any given ecosystem within the complexities of ecosystems will require these different actors to forge a shared agenda and worldview. The focus of their study was the Long-Term Social-Ecological Research Platform (LTSER) in the Cairngorms National Park (CNP; Scotland). Using semi-structured interviews of land-managers and scientists, a conceptual mapping of the CNP social–ecological system was created. It found that there were marked differences in interests between the two groups. Land-managers were found to be predominantly concerned by local-scale economic problems, while scientists were more concerned by environmental and global questions. However, both groups shared a common sense of uncertainty about the future which promoted a willingness for both groups to work together.

This volume present a snapshot of some of the work currently being done on the resilience of complex socioecological systems. As Martin et al. (Chapter 3) note, socioecological resilience research is still embryonic, but there is clearly plenty of exciting and challenging work remaining to be done. This thematic issue illustrates some of the most important steps being taken towards developing the approaches necessary to achieve a more sustainable future.

David A. Bohan

Alex J. Dumbrell

Acknowledgement

David A. Bohan would like to acknowledge the support of the FACCE SURPLUS project *PREAR* (ANR-15-SUSF-0002-03).

References

Holling, C.S., 1973. Resilience and stability of ecological systems. Annu. Rev. Ecol. Syst. 4, 1–23. https://doi.org/10.1146/annurev.es.04.110173.000245.

Holling, C.S., 1996. Engineering resilience versus ecological resilience. In: Engineering Within Ecological Constraints. National Academy Press, Washington, DC, pp. 31–43.

CHAPTER ONE

Adaptive capacity in ecosystems

David G. Angeler[a,b,*], **Hannah B. Fried-Petersen**[a], **Craig R. Allen**[c],
Ahjond Garmestani[d], **Dirac Twidwell**[e], **Wen-Ching Chuang**[d],
Victoria M. Donovan[e], **Tarsha Eason**[f], **Caleb P. Roberts**[e],
Shana M. Sundstrom[b], **Carissa L. Wonkka**[e]

[a]Department of Aquatic Sciences and Assessment, Swedish University of Agricultural Sciences, Uppsala, Sweden
[b]School of Natural Resources, University of Nebraska—Lincoln, Lincoln, NE, United States
[c]U.S. Geological Survey, Nebraska Cooperative Fish and Wildlife Research Unit, School of Natural Resources, University of Nebraska—Lincoln, Lincoln, NE, United States
[d]National Risk Management Research Laboratory, U.S. Environmental Protection Agency, Cincinnati, OH, United States
[e]Department of Agronomy & Horticulture, University of Nebraska, Keim Hall, Lincoln, NE, United States
[f]National Risk Management Research Laboratory, U.S. Environmental Protection Agency, Research Triangle Park, NC, United States
*Corresponding author: e-mail address: david.angeler@slu.se

Contents

Abstract

Understanding the capacity of ecosystems to adapt and to cope (i.e. adaptive capacity) with change is crucial to their management. However, definitions of adaptive capacity are often unclear and confusing, making application of this concept difficult. In this paper, we revisit definitions of adaptive capacity and operationalize the concept. We define adaptive capacity as the latent potential of an ecosystem to alter resilience in response to change. We present testable hypotheses to evaluate complementary attributes of adaptive capacity that may help further clarify the components and relevance of the concept. We suggest how sampling, inference and modelling can reduce key

Advances in Ecological Research, Volume 60
ISSN 0065-2504
https://doi.org/10.1016/bs.aecr.2019.02.001

uncertainties incrementally over time and increase learning about adaptive capacity. Improved quantitative assessments of adaptive capacity are needed because of the high uncertainty about global change and its potential effect on the capacity of ecosystems to adapt to social and ecological change. An improved understanding of adaptive capacity might ultimately allow for more efficient and targeted management.

1. Introduction

Environmental sustainability requires research that integrates human-nature interactions with sustainable practices to maintain resilient societal, cultural and economic requirements (triple bottom line; Lederwasch and Mukheibir, 2013); for instance, to foster ecosystem regimes that provide goods (e.g. food, fiber) and services (e.g. clean water, recreational activities) for humans (Kates et al., 2011). Ecosystems are subject to stresses (such as increasing intensification of agriculture, increasingly over-appropriated water supplies, and climate change), and these stresses can eventually surpass the planetary boundaries of sustainable use of natural resources (Rockström et al., 2009). The ability of ecosystems to adapt to these changes is limited as lakes shifting from clear-water to turbid regimes exemplify (Scheffer and van Nes, 2007). Eventually, ecosystems may undergo *regime shifts* (for definition of terms in italics see Box 1) to alternate species assemblages and ecosystem functioning at local, regional, and even global scales when their ability to adapt to stress becomes exhausted (Hughes et al., 2013). In this case, rather than adapting to prevailing abiotic and biotic conditions the system undergoes a profound and frequently irreversible transformation. The outcomes of regime shifts are highly uncertain but potentially have substantial negative effects on human health, security and welfare (Horner-Dixon, 1991; McMichael et al., 2008). This indicates the importance to proactively manage ecosystems in ways to increase their adaptive capacity and minimize the risk of regime shifts. Proactive management is important because reactive management (i.e. restoration) after regime shifts have occurred can be very costly, ineffective, uncertain and untenable in the long-term (Gulati et al., 2008; Palmer et al., 2010; Suding et al., 2004).

The concept of *adaptive capacity* has been rapidly assimilated in the social sciences and transdisciplinary social-ecological research (Folke et al., 2003; Gunderson, 2000), with multiple attempts made to formalize its meaning. Adaptive capacity is related to resilience theory (Gunderson and Holling, 2002; Holling, 1973), which has taken centre stage in the effort to

BOX 1

Definitions

Adaptive capacity: Latent property of an ecological system (or other complex system) to respond to disturbances in a manner that maintains the system within its current basin of attraction by altering the depth and/or width of that basin (Fig. 1)

Cross-scale Resilience: The degree to which a system has high functional diversity and high functional redundancy within and across the scales of the system

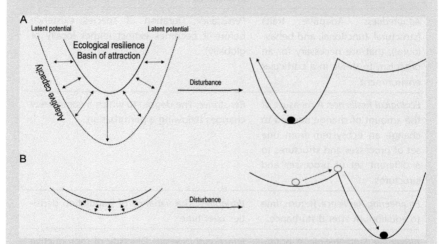

Fig. 1 Schematics illustrating high (A) and low (B) adaptive capacity. Adaptive capacity as a latent potential is shown by the dotted lines. Once expressed it triggers ecological resilience (lengths of arrows) and shapes the basin of attraction (left drawings). Drawings on the right show how high and low adaptive capacity influence the ability of a system to stay within its current regime (A) or transform to a new regime (B) after a disturbance.

Adaptation: Alterations in the structure or function of an organism due to natural selection by which the organism becomes better fitted to survive and reproduce in its environment

Ecological Memory: The collective representation of functional and structural attributes in an ecosystem that has been shaped by the systems' disturbance history

Continued

BOX 1—CONT'D

Adaptability: Ability to become adapted to live and reproduce under a particular range of environmental conditions

Stability: A system characteristic whereby system properties remain unchanged following disturbance. Adaptive capacity can increase stability, but system components can fluctuate (therefore being unstable) while still remaining within the range of values that signify a particular state. Components of *ecological stability* that influence disturbance responses are variability, resistance, engineering resilience and persistence

Adaptedness: Adaptive traits (structural, functional, and behavioural), that are necessary for an organism to thrive in a particular environment

Persistence: Duration of species existence before it becomes extinct (either locally or globally)

Ecological Resilience: A measure of the amount of change needed to change an ecosystem from one set of processes and structures to a different set of processes and structures

Resistance: The degree to which a community changes following a perturbation

Engineering Resilience: Return time to equilibrium after disturbance

Variability: The variance of population densities over time

Alternative State/Regime: A potential alternate configuration in terms of the structural and functional composition, processes, and feedbacks of a system

Functional Diversity: Diversity of reproductive phenology, seed bank potential, colonization and dispersal abilities, and other traits. This can enhance adaptive capacity by increasing functional redundancy and response diversity (see below)

Basin of Attraction (stability domain): A region of the state space where the system tends to remain and has a definable configuration in terms of the abundance, composition, and processes of a system

Functional Redundancy: Existence of more than one species or process delivering the same ecological function. This contributes to adaptive capacity in ecosystems by providing buffering for loss of function due to disturbance-induced mortality

Response Diversity: Variability among individuals or species in the range of response patterns to disturbances

Regime Shift: Persistent change in structure, function, and feedbacks of an ecosystem

understand ecosystem dynamics during change. The concept of adaptive capacity has, in parallel with the transdisciplinary development of resilience theory, helped to diversify the meanings and definitions of systems undergoing change (Gallopín, 2006). Adaptive capacity has been mainly used qualitatively in climate change, vulnerability, and a risk/disaster management context in the social sciences and varies between different contexts and systems (Adger et al., 2007). Similarly, in the ecological sciences, *adaptation, adaptedness, adaptability* and *adaptive capacity*, terms with different meanings, have often been used interchangeably (Gallopín, 2006; Smit and Wandel, 2006). Consequently, operationalizing the concept of adaptive capacity, and by extension resilience theory for application and management, has been difficult, because of a loss of clarity and loose, incorrect and often normative use of these disparate concepts (Angeler and Allen, 2016; Brand and Jax, 2007). Misuse of terms can have significant negative impacts because resilience and adaptive capacity are being used to help guide responses to extreme events. Furthermore, assessments of ecosystems that drive international research priorities depend on a comprehensive understanding of these concepts (Smit and Wandel, 2006).

Because the concept of adaptive capacity is muddied with multiple meanings, its current use often makes it indistinguishable from resilience. In this paper, our goal is to clearly define the concept of adaptive capacity in ecosystems with the aim of differentiating it from similar concepts, particularly *ecological resilience* (Angeler and Allen, 2016) and *ecological stability* (Donohue et al., 2016) (Box 1). Since approaches for operationalization and quantification of these concepts are needed, we describe components of adaptive capacity in ecosystems and discuss how they might mitigate and direct ecological response to ongoing environmental change. Further, we identify a research agenda to test hypotheses related to adaptive capacity and the ability of ecosystems to cope with environmental change.

2. Definitions and formalization

Much of the terminology and definitions used in the ecological adaptive capacity context has a Darwinian adaptation focus on species. This is reflected in the currently most comprehensive definition of adaptive capacity (Beever et al., 2015; Nicotra et al., 2015): These authors define adaptive capacity for species, but also populations, as a combination of evolutionary potential, dispersal ability, life-history traits and phenotypic plasticity, which are influenced by genetic, epigenetic, behavioural and acclimation processes.

This definition is well aligned with the broad use of the term adaptation in ecology, which is defined as an organism's ability to cope with environmental changes in order to survive and reproduce (Smit and Wandel, 2006). The term adaptation itself is often used interchangeably with the term *adaptability* or *acclimation*, which, as used in biology, considers the ability of a species to become adjusted and to live and reproduce under a certain range of environmental conditions (Bowler, 2005; Conrad, 1983). Acclimation and adaption processes are relevant from an evolutionary point of view because the long-term persistence of species in an ecosystem could be governed by either the ability to adapt or to acclimate depending on the degree of perturbations being experienced and whether these are within the range of those species' abilities to acclimate or not. Another term, *adaptedness*, has a more specific meaning than adaptation or adaptability. Dobzhansky (1968) defined adaptedness as the adaptive traits (structure, function, and behaviour of an organism) that are crucial for an organism to thrive in an environment. Adaptedness embraces species-specific adaptation to a certain range of environmental conditions. Adaptedness is therefore context dependent and not a generic property as adaptability or adaptation would suggest. That is, high adaptedness does not necessarily mean high adaptability because a species may be highly adapted to a special and constant environment but have little capacity to adapt to other environments or to changes in its environment (Gallopín, 2006). For example, a cold-stenothermic mayfly may thrive in an arctic stream, but it does not have the necessary adaptation to live in tropical lakes or to keep pace with warming of the arctic stream environment. Adaptedness can be tested through reciprocal transplant experiments to assess phenotypic fitness to local ecological niches.

Despite the dominant focus of adaptive capacity on lower levels of biological organization in the literature, the term is increasingly used as an ecosystem property, implicitly recognizing that adaptation emerges from such lower biological levels. In an ecosystem context, adaptive capacity recognizes that the ability of ecosystems to cope with disturbances is limited and that regime shifts can occur. Gunderson (2000) defined adaptive capacity as a system property, where adaptive capacity modifies ecological resilience (or "*basin of attraction*"). This definition is very similar to the original definition of ecological resilience: "Resilience is a measure of the persistence of systems and of their ability to absorb change and disturbance and still maintain the same relationships between populations or state variables" (Holling, 1973). It follows that both Holling's and Gunderson's definition focus on a single regime rather than multiple alternative ecosystem regimes. This single equilibrium view is also inherent in the multidimensional components of ecological stability (Donohue et al., 2013, 2016).

Inherent to ecological resilience is the capacity of ecosystems to undergo regime shifts, meaning that ecosystems can exist in more than one regime. Gunderson and Holling (2002) defined ecological resilience as "the magnitude of disturbance that can be absorbed before the system changes its structure by changing the variables and processes that control behavior". Similarly, in a recent overview of resilience definitions, Angeler and Allen (2016) refer to ecological resilience as "a measure of the amount of change needed to change an ecosystem from one set of processes and structures to a different set of processes and structures."

Ecological resilience encompasses broader system dynamics by considering both adaptation within, and shifts between, alternative basins of attraction (i.e. *alternative regimes*) (e.g. shifts from clear-water to turbid regimes of shallow lakes). The distinction between single vs multiple regimes helps distinguish adaptive capacity from ecological resilience. Adaptive capacity focuses on dynamics within a specific regime (e.g. disturbance-response patterns that are distinctly different between these two lake regimes), and therefore adaptive capacity and ecological stability are subsets of ecological resilience. Ecological resilience is explicitly concerned with dynamics both within and between regimes. Thus, similar to the view of Gunderson (2000), ecosystem adaptive capacity can be formalized and defined as follows (Fig. 1, Box 1): *Adaptive capacity is the latent potential of an ecosystem to alter resilience in response to change.* In particular, adaptive capacity is a dynamic process that contributes to mold a specific regime state of an ecosystem in response to disturbances. Our emphasis on latent is key to how we differentiate adaptive capacity from ecological resilience. The components that drive adaptive capacity are present but not always observable until disturbance triggers their manifestation. When one or multiple components of adaptive capacity become active, resilience is manifested. This means that an ecosystem can show trajectories, which either keep them in the same regime or move them to an alternative regime (Fig. 1).

The consideration of adaptive capacity as a subset of ecological resilience has applied relevance. Research is increasingly focused on the assessment of

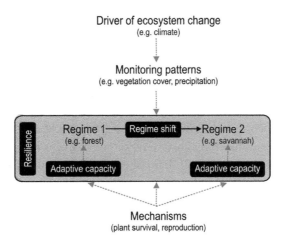

Fig. 2 Schematic showing the relationship between drivers that cause regime shifts in ecosystems and associated mechanisms responsible for adaptive capacity and resilience. Monitoring can be based on patterns of system change. Adding an assessment of mechanisms can deepen our understanding of the link between adaptive capacity, ecological resilience, and likelihood of regime shifts.

early warning signals of impending regime shifts (e.g. Dakos et al., 2015; Spanbauer et al., 2014, 2016), with the goal of employing management intervention if appreciable signals are detected. Although such studies implicitly consider an exhaustion of adaptive capacity and a reduction in resilience that shrink the basin of attraction, the underlying mechanisms are not fully accounted for (Fig. 2). Consider an Amazon vegetation model, where aridity drives system movement between a forest regime and a savannah regime (Mitra et al., 2015). The change in precipitation is the driver, but the actual mechanisms have to do with plant survival and reproduction when stressed by increased aridity. Similarly, in shallow lakes nutrients drive shifts between clear-water and turbid regimes but mechanisms are related to, for instance, changes in oxygen concentrations and habitat complexity (Scheffer, 1997; Scheffer and van Nes, 2007). That these mechanisms do not need to be measured in order to calculate an early warning signal, which could use change in percent cover of plants or variance in precipitation, is a strength of such signals. But it does not negate the value of a deeper understanding of the mechanistic link between adaptive capacity, ecological resilience, and likelihood of regime shifts (Clements and Ozgul, 2018). In particular, it would be useful to understand mechanistically the system features from which adaptive capacity derives and how particular disturbances may impact these various components of adaptive capacity. We discuss components of adaptive capacity and forward hypotheses to test the generalized adaptive capacity of a system. Our hypotheses suggest ways in which monitoring can be used to assess how adaptive a system is to disturbances.

3. Components of adaptive capacity

Adaptive capacity as a latent potential of ecosystems is comprised of components that are dynamically interlinked so only partially lend themselves to being organized in discrete categories (Table 1). Among these components are ecological memory, cross-scale interactions, ecological functioning, and rare species.

3.1 Ecological memory and legacy

The composition and distribution of organisms, their interactions in space and time, and their life-history experience with environmental fluctuations contribute to *ecological memory* (Nyström and Folke, 2001). Ecological memory has been defined as "the capability of the past states or experiences of a community to influence the present or future ecological responses of the community" (Sun and Hai, 2011). Specifically, ecological memory comprises all structural and functional features of ecological communities, which have been shaped by the interaction of past disturbances (natural and anthropogenic), spatial aspects (dispersal, habitat connectivity), biological interactions (competition, predation), phenotypic plasticity, evolutionary (speciation, extinctions, anagenesis, random mutations) and phylogenetic processes. A related term, disturbance legacy, focuses on species traits and adaptations that contain information about successful strategies to past disturbances, as well as residual organisms, propagules, and physical structures arising from past biological activity (Johnstone et al., 2016). For example, shade-tolerant understory trees with large seedling banks can take rapid advantage of canopy gaps after windstorms. The evolution of adaptive disturbance-response traits provides a competitive advantage and is a form of ecological memory, where past windstorm regimes selected for post-windstorm recruitment advantages (Johnstone et al., 2016).

This memory of ecological communities allows for a "learning process" (Carpenter et al., 2001) that facilitates persistence during future disturbance. Power et al. (2015) use learning theory to robustly demonstrate that communities can change composition to facilitate persistence. This is due to individual selection acting upon interspecific correlations, which create a powerful "distributed ecological memory of multiple past regimes that

Table 1 Factors that contribute to adaptive capacity across different scales of biological organization.
Hierarchy of organization and selected traits

- Sub-individual
  - Matching physiological conditions to fluctuating inputs or internal demands (allostasis) (Carpenter and Brock, 2008)
  - Genetic, epigenetic and molecular processes (e.g. mutation)
- Individual
  - Phenotypic plasticity
  - Learning and dispersal ability
  - Behaviour
  - Adaptive evolution related to genetic diversity and evolutionary rates
  - Links between life-history traits, phenotypic plasticity, and evolutionary potential
- Population
  - Heritable life history characteristics: generation time, reproductive capacity, migration, habitat selection, genome size, survival characteristics (resting stages; hibernation, estivation), generalist vs specialist species
  - Population structure
  - Metapopulation dynamics
- Community
  - Taxonomic diversity
  - Functional diversity (redundancy, response diversity)
  - Strength of species interactions
  - Metacommunity dynamics (colonization and dispersal abilities)
  - Founder effects
  - Priority effects
  - Dormancy (resting eggs and propagule banks) and bet-hedging strategies
- Ecosystem
  - Interaction of and connection between abiotic and biotic elements in feedback loops (balancing and reinforcing or negative and positive)
  - Changing shapes of basin of attraction/stability landscape (topography, soils, landforms)
- Biome
  - Biogeographical distributions of native and invasive species
  - Phylogenetic dynamics
  - Evolutionary, disturbance and climate histories

Note: this table is not exhaustive and meant only to highlight the complexity of factors influencing adaptive capacity.

allows a community to accurately restore historical compositions from small fragments; recover a regime composition following disturbance; and correctly classify ambiguous initial compositions" (i.e. community rescue; Low-Décarie et al., 2015). Adaptive capacity explicitly accounts for pattern-process relationships of ecological memory that operate within and across the hierarchy of biological organization (i.e. they contribute to *cross-scale resilience*) (Table 1).

3.2 Cross-scale interactions

Ecosystems are hierarchically organized and have distinct patterns of structure, function, and processes that are compartmentalized by spatiotemporal scales. The consideration of cross-scale interactions is important because the impact of disturbance in ecosystems can be scale-specific (Nash et al., 2014; Pickett and White, 1985). That is, if disturbances affect components of ecological memory or other features of adaptive capacity at one scale, other components at other scales might buffer the disturbances and maintain system-level functioning and resilience via processes of compensation, behavioural plasticity, or increasing abundance (Allen and Holling, 2008; Allen et al., 2005; Peterson et al., 1998). These cross-scale buffering processes are considered critical to adaptive capacity as they diminish the impact of the disturbance on system-level functioning so are crucial for maintaining a system in a regime. This buffering ability can be further explored through assessments of functional ecosystem characteristics, such as the promotion of species coexistence through trait adaptation (Klauschies et al., 2016).

3.3 Ecological functioning

Ecosystem reactions to disturbances, including buffering responses that dampen the impact of the disturbance, rely on functional responses to perturbations, which in turn depends on the diversity of expressible traits such as reproductive phenology, seed bank potential, colonization and dispersal abilities (i.e. *functional diversity*) that provide a range of response patterns to disturbances (i.e. *response diversity*) (Elmqvist et al., 2003). A recent study on coral reefs (Nash et al., 2016) and a meta-analysis of forest resilience (Cole et al., 2014) support the importance of response diversity and cross-scale resilience for recovery after disturbances. Additionally, response and effect trait patterns and ecological network structure influence response diversity and ecosystem service provisioning (Mori et al., 2013; Oliver et al., 2015; Schleuning et al., 2015). In addition to diversity, redundancies of functional traits such as functional feeding groups of animals (predators, collectors) or seed mass and canopy height in plants (*functional redundancy*) are important for the stabilization of system-level processes (e.g. primary production, decomposition) and feedbacks, and therefore contribute to the adaptive capacity of an ecosystem. Assessing the distribution, diversity, and redundancy of functional traits within and across spatiotemporal scales may be one possible measurable surrogate for some key components of adaptive capacity (Table 1), and may provide an indicator of the erosion of adaptive capacity as a result of environmental change (Laliberte et al., 2010).

3.4 Rare species

Rare species are increasingly recognized to play critical roles in ecosystem (Lyons et al., 2005). Their abundances can vary with the scales of observation (within and across ecosystems), as is the case with biodiversity in general (Angeler and Drakare, 2013), and their functional contribution to ecosystem processes can differ from those of dominant species (Hua et al., 2015). Based on current evidence rare species might play different roles in the adaptive capacity of ecosystems. First, the importance of rare species is evident in their ability to replace dominant species after perturbation and maintain ecological functions in the system, which in turn contributes to adaptive capacity (Walker et al., 1999). For instance, rare shrub species with larger root crowns than dominant species were able to compensate for the loss of dominant shrub species to mechanical disturbance by re-sprouting prolifically, thus maintaining a shrub-dominated system despite disturbance (Wonkka et al., 2016). This example suggests that rare species can contribute to adaptive capacity in ecosystems. However, recent research also shows the contrary. Mouillot et al. (2013) found that rare species in alpine meadows, coral reefs, and tropical forests supported functional trait combinations that were not represented by abundant species. If rare species go extinct with ongoing environmental change, negative effects on ecosystem processes may result from the subsequent loss of adaptive capacity. Such effects may occur even if biodiversity associated with abundant species is high (Mouillot et al., 2013).

These examples show that rare species may contribute an important but, to some extent unpredictable degree of adaptive capacity to ecosystem change. This highlights the need for further research to understand basic structural (richness) and functional (traits) aspects of rare species, and the interaction between them (Levins and Culver, 1971), across ecosystem types (from low to high diversity systems) and their role in conferring adaptive capacity (Angeler et al., 2015).

4. Assessing adaptive capacity

Assessing adaptive capacity requires methods to integrate the hierarchical, scaled structure of systems with the many features and components discussed previously. Some components of adaptive capacity are contained in one level of organization, such as genetic diversity within a species, while for others, such as the buffering processes discussed, the essence of the adaptive capacity lies in the ability of a scale-specific disturbance to be buffered by compensatory processes operating at unaffected scales. The objective identification of relevant scales becomes a necessary component of any adaptive capacity assessment. Previous resilience assessments have used discontinuity approaches to objectively identify the scaling structure present in ecosystems (Angeler et al., 2016a). The discontinuity analyses have so far shown promising results in assessing resilience of aquatic and terrestrial ecosystems (Angeler et al., 2016b) as well as in other complex systems (economic, anthropological, social–ecological; Garmestani et al., 2009; Sundstrom et al., 2014). Discontinuity analysis may therefore also be useful in assessing the adaptive capacity of ecosystem regimes.

Because information about ecosystems is frequently limited, adaptive capacity can be assessed following a recently proposed hypothesis-testing framework for quantifying ecological resilience (Baho et al., 2017). This evaluation comprises initial assessments of specific facets of adaptive capacity and then tests and recalibrates hypotheses iteratively to increase knowledge and provide learning opportunities about the ecosystem's general adaptive capacity.

Surrogates of adaptive capacity can be evaluated measuring simple indicators of ecological stability (*resistance, persistence, variability,* and *engineering resilience*) (Donohue et al., 2013), biodiversity (Magurran, 2004), and resilience (Angeler et al., 2016b). The stability aspects can be evaluated through repeated sampling and quantification of structural and functional variables (e.g. diversity, abundance, evenness, community composition, functional redundancies, and diversity and process rates) within and across scales.

The initial step for quantifying adaptive capacity builds on Carpenter et al. (2001) to test for the "adaptive capacity of what to what". However, testing for specific aspects of adaptive capacity may not be representative of the general adaptive capacity of an ecosystem. This is because there is limited surrogacy of metrics when assessing ecological responses to stressors (Johnson and Hering, 2009). In addition, focusing on specified adaptive capacity can be problematic because managing adaptive capacity of particular parts of an ecosystem, especially in terms of managing for predictable outcomes of disturbances or provision of ecosystem services, may cause the system to lose adaptive capacity or resilience in other ways (Carpenter et al., 2015). That is, if the focus of monitoring is on specific components of adaptive capacity, then adaptive capacity as a whole is neglected. Specified assessments of adaptive capacity shall therefore be regarded as an initial step towards assessing the broader systemic or general adaptive capacity of an ecosystem. It follows that assessing and managing for general adaptive capacity will require the simultaneous assessment of a range of variables to cover generic system properties and create possibilities for integral, resilience-based ecosystem management.

4.1 Hypothesis testing to clarify adaptive capacity concepts

We suggest that these difficulties can be overcome by implementing adaptive monitoring and management. For this purpose, posing hypotheses that test premises of adaptive capacity are helpful (Table 2). We propose hypotheses that are not mutually exclusive and are well aligned with our adaptive capacity definition, allowing for the evaluation of its attributes in a logical, iterative sequence. These hypotheses can be tested using quantifiable stability and resilience measures (Angeler et al., 2016a) based on multiple lines of evidence (e.g. taxa across distinct trophic levels; Burthe et al., 2016). Most hypotheses can be framed specifically from a management perspective to facilitate the quantification of adaptive capacity without sacrificing the complexity inherent in management-related assessments. Also, most of our proposed hypotheses are supported by empirical observations (examples in Table 2), suggesting implementation of our quantification framework with ecological realism.

Adaptive management can start by choosing species that might serve as sentinels of system change (e.g. species sensitive to global warming) (Angeler et al., 2016a). Next, sampling can be adapted to select appropriate spatial and/or temporal scales for monitoring to account for the cross-scale structure present in the system (Thompson and Seber, 1996). This can contribute to pattern identification following population responses of sentinel species to disturbances. Incorporation of genetic, evolutionary, molecular, and physiological variables and the measurement of process rates in monitoring can increase inference about ecosystem change, providing information for recalibration of management hypotheses. Monitoring can be refined by

Table 2 Premises of adaptive capacity components and simple, management-relevant, mutually non-exclusive hypotheses.

	Adaptive capacity	
	Premise 1 (system regime stable)	Premise 2 (system regime erodes)
Hypotheses		
1. Adaptive capacity surrogates	High persistence, resistance and recovery; low variability	Slowed down recovery, decreasing resistance and persistence, and higher variability
2. Ecological memory	High structural and functional redundancy and diversity; high rare trait combinations; persistence of species with evolved disturbance response traits; nearby reservoirs of propogules, seeds, larvae, and individuals that can repopulate nearby disturbed areas	Decreasing redundancy and diversity and rare trait combinations
3. Keystone/ deterministic/ stochastic species	Contribute to ecological memory	Increasing extinctions
4. Response diversity	High	Reduced
5. Functional compensation	High	Decreasing
6. Functional trait distributions	High diversity and redundancy within and across scales	Decreasing diversity and redundancy within and across scales
Support	Allen et al. (2005); Angeler et al. (2016a); Bellingham et al. (1995); Baho et al. (2014); Boucher et al. (1994); Kühsel and Blüthgen (2015); Mouillot et al. (2013); Nash et al. (2016); Walker et al. (1999)	Carpenter and Brock (2008); Dakos et al. (2008); Hooper et al. (2012); Mumby et al. (2014); Nyström (2006); Nash et al. (2016); Spanbauer et al. (2016)
Management implication	Maintain regime	(1) Adaptive management and governance to stave off regime shifts (2) Scenario planning and transformative governance if adaptive capacity gets exhausted

These serve as a starting point for reiterative hypothesis testing and can be refined, modified and adapted to specific ecosystems during monitoring.

subsequently recalibrating hypotheses in an adaptive process. This is necessary because measurements and monitoring at inappropriate scales may miss relevant patterns and processes in the system (identifying false negatives; Type II error); for instance, when sampling designs fail to detect species in an ecosystem. Once Type II errors are reduced and appropriate scales identified, the estimation of species abundances and occurrences in the ecosystem may be highly uncertain due to spatial and temporal variability (identifying false positive; Type I error). However, increasing knowledge of the system through continued monitoring and learning can reduce uncertainty. Thus, adaptive management first focuses on reducing Type II errors sufficiently such that subsequent analyses can focus on the reduction of uncertainty (adaptive inference, Holling and Allen, 2002). Type II errors can be reduced by assessing adaptive capacity attributes (e.g. cross-scale and within scale structure and associated functional diversity redundancy) when ecological information of the ecosystem is limited. This can be done with the analysis of temporal snapshots, which are often the only resource available to managers. Subsequently, monitoring can be designed, implemented, and sequentially modified to successively reduce Type I errors; that is, by improving knowledge of a broader range of adaptive capacity characteristics that need to be sampled over time (e.g. how fast is recovery after a disturbance). Such recalibrations can target functional assessments of sentinel species to change and, in further iterations, be extended to other taxa. This type of hypothesis testing builds on adaptive management (Allen et al., 2011), sampling (Thompson and Seber, 1996), modelling (Uden et al., 2015), and inference (Holling and Allen, 2002). It allows revealing, refining, understanding, and ultimately managing general ecosystem adaptive capacity, while increasing learning and reducing uncertainty. In this process, experiments can be designed that sequentially recalibrate strategies based on the outcomes of previous experiments and from which decisions about further data generation and monitoring can be made (Fig. 3).

5. Managing adaptive capacity

Our suggested hypotheses are very general at this stage, but they can provide an initial step to inform management. First, managing for adaptive capacity may help maintain ecosystems in regimes that sustain human livelihoods (Allen et al., 2011). In this case maintaining adaptive capacity is fundamental for managing ecosystems away from critical thresholds upon which a system shifts into a novel regime. Crucial to managing for adaptive capacity

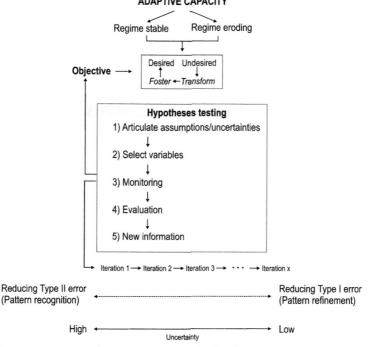

Fig. 3 Reiterative testing, recalibrating, and refining of explicit hypotheses of adaptive capacity within an adaptive management, inference, and modelling framework. The approach first recognizes patterns (reducing risk of type II error) and then refines knowledge about patterns reiteratively (reducing risk of type I error) to meet adaptive or transformative management objectives and reduce uncertainty.

is the consideration of cross-scale interactions across hierarchical levels and temporal scales. Verifying our premises while testing hypotheses iteratively can be useful for designing management interventions to foster the adaptive capacity of a specific desired regime. A combination of ecological and technological approaches might be necessary to this end (Rist et al., 2014). Some of these approaches, e.g., assisted translocations or introductions of species to compensate for lost crucial functions in an ecosystem (Chaffin et al., 2016a), are controversial and potentially limited in their application to management. Second, when an ecosystem fails to provide ecosystem goods and services, management can deliberately reduce adaptive capacity in order to induce a shift towards a more desired regime, and then foster the adaptive capacity of the new regime (Chaffin et al., 2016b) (Fig. 3). There is, for instance, a rich body of literature on lake biomanipulation, which exemplifies transformation of degraded lakes into more desirable systems (Hansson et al., 1998).

The biomanipulation example is useful because while it offers possible management options, it also highlights potential limitations when managing for adaptive capacity. Lake biomanipulation often consists of management interventions based on ecological (food web manipulations) and technological interventions (e.g. water column aeration, sediment dredging or lining, nutrient precipitation). However, lessons from biomanipulation have shown that these solutions can incur short- to long-term costs that may not be tenable for most systems.

It is clear that a series of ecological, resource and ethical issues may currently complicate the translation of a solid body of theory on adaptive capacity to its management on the ground. Current environmental policy further limits the implementation of resilience-based approaches to facilitate management (Green et al., 2015).

6. Conclusion

This paper suggests a way forward to enhance our ability to explicitly define and reduce uncertainties and promote more holistic and effective modelling, management, and monitoring of adaptive capacity. In addition to testing premises and hypotheses of adaptive capacity, defining highly-related concepts will aid in continuing towards operationalizing adaptive capacity. For instance, although we have touched on how adaptive capacity may contribute to the width and depth of the basin of attraction, methods for defining the actual basin of attraction are few and lack rigorous testing (Gunderson, 2000). It is also unclear how to estimate something that is not static, and is constantly experiencing feedback from manifested resilience. Despite these and other non-trivial challenges, assessing the basin of attraction would allow estimates of adaptive capacity to move beyond point measures comparable only between subsequent measures towards a direct estimate of adaptive capacity relative to the system's potential adaptive capacity (Carpenter et al., 2001). Defining the basin of attraction would also allow the estimation of the size of the buffer that the current level of adaptive capacity provides against system regime shifts and transformations, and it would allow transformative elements (e.g. invasive species) to be distinguished from elements that contribute to adaptive capacity (e.g. rare species; Elmqvist et al., 2003, Folke et al., 2010). However, new methods for detecting spatial regimes and discontinuous resource aggregations show promise for delineating basins of attraction in space and over time (Allen et al., 2016; Angeler et al., 2016b; Sundstrom et al., 2017).

Uncertainty can never be fully eliminated, but it can be reduced incrementally while an ecosystem is monitored, modelled and managed over time. Explicit learning during this process can overcome common management problems, such as delayed action under uncertainty (Conroy et al., 2011), prioritization of limited financial resources (Stewart-Koster et al., 2015), and the limited coordination in governance of natural resources (Cumming et al., 2013). Such improvements are needed because of uncertainty about global change impacts on an ecosystem's ability to absorb disturbances. An improved understanding of adaptive capacity can ultimately help to facilitate ecosystem management within current ecological, economic, and ethical constraints. Our approach to assessing adaptive capacity provides insight into the challenges of accounting for ecological complexity in ecosystem management. It particularly highlights enormous resource needs for practical implementation of adaptive capacity assessments, and pinpoints persistent challenges in closing gaps between science, policy, and management (Garmestani and Benson, 2013).

Acknowledgements

This study was funded through grants from the Swedish Research Councils FORMAS (2014-1193) and VR (2014-5828), and The United States Department of Defense, SERDP (RC-2510). The Nebraska Cooperative Fish and Wildlife Research Unit is jointly supported by a cooperative agreement between the U.S. Geological Survey, the Nebraska Game and Parks Commission, the University of Nebraska – Lincoln, the United States Fish and Wildlife Service and the Wildlife Management Institute. The views expressed in this paper are those of the authors and do not necessarily represent the views or policies of the U.S. Environmental Protection Agency. This research was performed while WC held NRC Research Associateship awards at the U.S. EPA.

References

Adger, W.N., Agrawala, S., Mirza, M.M.Q., Conde, C., O'Brien, K., Pulhin, J., Pulwarty, R., Smit, B., Takahashi, K., 2007. Assessment of adaptation practices, options, constraints and capacity. In: Parry, M.L., Canziani, O.F., Palutikof, J.P., van der Linden, P.J., Hanson, C.E. (Eds.), Climate Change 2007: Impacts, Adaptation and Vulnerability. Contribution of Working Group II to the Fourth Assessment Report of the Intergovernmental Panel on Climate Change. Cambridge University Press, Cambridge, UK, pp. 1–22.

Allen, C.R., Holling, C.S., 2008. Discontinuities in Ecosystems and Other Complex Systems. Columbia University Press, New York.

Allen, C.R., Gunderson, L., Johnson, A.R., 2005. The use of discontinuities and functional groups to assess relative resilience in complex systems. Ecosystems 8, 958–966.

Allen, C.R., Fontaine, J.J., Pope, K.L., Garmestani, A.S., 2011. Adaptive management for a turbulent future. J. Environ. Manage. 92, 1339–1345.

Allen, C.R., Angeler, D.G., Cumming, G.S., Folke, C., Twidwell, D., Uden, D.R., 2016. Quantifying spatial resilience. J. Appl. Ecol. 53, 625–635.

Angeler, D.G., Allen, C.R., 2016. Quantifying resilience. J. Appl. Ecol. 53, 617–624.

Angeler, D.G., Drakare, S., 2013. Tracing alpha, beta, and gamma diversity responses to environmental change in boreal lakes. Oecologia 172, 1191–1202.

Angeler, D.G., Allen, C.R., Uden, D.R., Johnson, R.K., 2015. Spatial patterns and functional redundancies in a changing boreal lake landscape. Ecosystems 18, 889–902.

Angeler, D.G., Allen, C.R., Barichievy, C., Eason, T., Garmestani, A.S., Graham, N.A.J., Granholm, D., Gunderson, L., Knutson, M., Nash, K.L., Nelson, R.J., Nyström, M., Spanbauer, T.E., Stow, C.A., Sundstrom, S.M., 2016a. Management applications of discontinuity theory. J. Appl. Ecol. 53, 688–698.

Angeler, D.G., Allen, C.R., Garmestani, A.S., Gunderson, L.H., Linkov, I., 2016b. Panarchy use in environmental science for risk and resilience planning. Environ. Syst. Decis. 36, 225–228.

Baho, D.L., Drakare, S., Johnson, R.K., Allen, C.R., Angeler, D.G., 2014. Similar resilience attributes in lakes with different management practices. PLoS One 9 (3), e91881. https://doi.org/10.1371/journal.pone.0091881.

Baho, D.L., Allen, C.R., Garmestani, A., Fried-Petersen, H.B., Renes, S.E., Gunderson, L.H., Angeler, D.G., 2017. A quantitative framework for assessing ecological resilience. Ecol. Soc. 22 (3), 17. https://doi.org/10.5751/ES-09427-220317.

Beever, E.A., O'Leary, J., Mengelt, C., West, J.M., Julius, S., Green, N., Magness, D., Petes, L., Stein, B., Nicotra, A.B., Hellmann, J.J., Robertson, A.L., Staudinger, M.D., Rosenberg, A.A., Babij, E., Brennan, J., Schuurman, G.W., Hofmann, G.E., 2015. Improving conservation outcomes with a new paradigm for understanding species' fundamental and realized adaptive capacity. Conserv. Lett. 9, 131–137.

Bellingham, P.J., Tanner, E.V.J., Healy, J.R., 1995. Damage and responsiveness of Jamaican montane tree species after disturbance by a hurricane. Ecology 76, 2562–2580.

Boucher, D.H., Vandermeer, J.H., Mallona, M.A., Zamora, N., Perfecto, I., 1994. Resistance and resilience in a directly regenerating rainforest: Nicaraguan trees of the Vochysiaceae after Hurricane Joan. For. Ecol. Manage. 68, 127–136.

Bowler, K., 2005. Acclimation, heat shock and hardening. J. Therm. Biol. 30, 125–130.

Brand, F.S., Jax, L., 2007. Focusing the meaning(s) of resilience: resilience as a descriptive concept and a boundary object. Ecol. Soc. 12 (1), 23.

Burthe, S.J., Henrys, P.A., Mackay, E.B., Spears, B.M., Campbell, R., Carvalho, L., Dudley, B., Gunn, I.D.M., Johns, D.G., Maberly, S.C., May, L., Newell, M.A., Wanless, S., Winfield, I.J., Thackeray, S.J., Daunt, F., 2016. Do early warning indicators consistently predict nonlinear change in long-term ecological data? J. Appl. Ecol. 53, 666–676.

Carpenter, S.R., Brock, W.A., 2008. Adaptive capacity and traps. Ecol. Soc. 13 (2), 40. [online] URL: http://www.ecologyandsociety.org/vol13/iss2/art40/.

Carpenter, S., Walker, B., Anderies, J.M., Abel, N., 2001. From metaphor to measurement: resilience of what to what? Ecosystems 4, 765–781.

Carpenter, S.R., Brock, W.A., Folke, C., van Nes, E.H., Scheffer, M., 2015. Allowing variance may enlarge the safe operating space for exploited ecosystems. Proc. Natl. Acad. Sci. U. S. A. 112 (46), 14384–14389.

Chaffin, B.C., Garmestani, A.S., Gunderson, L., Harm Benson, M., Angeler, D.G., Arnold, C.A., Cosens, B., Craig, R.K., Ruhl, J.B., Allen, C.R., 2016a. Transformative environmental governance. Annu. Rev. Env. Resour. 41, 399–423.

Chaffin, B.C., Garmestani, A.S., Angeler, D.G., Herrmann, D.L., Hopton, M.E., Kolasa, J., Nyström, M., Sendzimir, J., Stow, C.A., Allen, C.R., 2016b. Adaptive governance, biological invasions and ecological resilience. J. Environ. Manage. 183, 399–407. https://doi.org/10.1016/j.jenvman.2016.04.040.

Clements, C.F., Ozgul, A., 2018. Indicators of transitions in biological systems. Ecol. Lett. 21, 905–919. https://doi.org/10.1111/ele.12948.

Cole, L.E., Bhagwat, S.A., Willis, K.J., 2014. Recovery and resilience of tropical forests after disturbance. Nat. Commun. 5, 3906.

Conrad, M., 1983. Adaptability, the Significance of Variability From Molecule to Ecosystem. Plenum Press, New York.

Conroy, M.J., Runge, M.C., Nichols, J.D., Stodola, K.W., Cooper, R.J., 2011. Conservation in the face of climate change: the role of alternative models, monitoring, and adaptation in confronting and reducing uncertainty. Biol. Conserv. 144, 1204–1213.

Cumming, G.S., Olsson, P., Chapin III, F.S., Holling, C.S., 2013. Resilience, experimentation, and scale mismatches in social-ecological landscapes. Landsc. Ecol. 28, 1139–1150.

Dakos, V., Scheffer, M., van Nes, E.H., Brovkin, V., Petoukhov, V., Held, H., 2008. Slowing down as an early warning signal for abrupt climate change. Proc. Natl. Acad. Sci. U. S. A. 105 (38), 14308–14312.

Dakos, V., Carpenter, S.R., van Nes, E.H., Scheffer, M., 2015. Resilience indicators: prospects and limitations for early warnings of regime shifts. Phil. Trans. R. Soc. B 370, 1–10.

Dobzhansky, T., 1968. On Some Fundamental Concepts of Darwinian Biology. In: Dobzhansky, T., Hecht, M.K., Steere, W.C. (Eds.), Evolutionary Biology. Plenum Press, New York, pp. 1–34.

Donohue, I., Petchey, O.L., Montoya, J.M., Jackson, A.L., McNally, L., Viana, M., Healy, K., Lurgi, M., O'Connor, N.E., Emmerson, M.C., 2013. On the dimensionality of ecological stability. Ecol. Lett. 16, 421–429.

Donohue, I., Hillebrand, H., Montoya, J.M., Petchey, O.L., Pimm, S.L., Fowler, M.S., Healy, K., Jackson, A.L., Lurgi, M., McClean, D., O'Connor, N.E., O'Gorman, E.J., Yang, Q., 2016. Navigating the complexity of ecological stability. Ecol. Lett. 19, 1172–1185.

Elmqvist, T., Folke, C., Nyström, M., Peterson, G., Bengtsson, J., Walker, B., Norberg, J., 2003. Response diversity, ecosystem change and resilience. Front. Ecol. Environ. 1, 488–494.

Folke, C., Colding, J., Berkes, F., 2003. Synthesis: building resilience and adaptive capacity in social-ecological systems. In: Navigating Social-Ecological Systems: Building Resilience for Complexity and Change. vol. 9. Cambridge University Press, pp. 352–387. 1.

Folke, C., Carpenter, S.R., Walker, B., Scheffer, M., Chapin, T., Rockström, J., 2010. Resilience thinking: integrating resilience, adaptability and transformability. Ecol. Soc. 15 (4), 20.

Gallopín, G.C., 2006. Linkages between vulnerability, resilience, and adaptive capacity. Glob. Environ. Chang. 16, 293–303.

Garmestani, A.S., Benson, M.H., 2013. A framework for resilience-based governance of social-ecological systems. Ecol. Soc. 18 (1), 9. https://doi.org/10.5751/ES-05180-180109.

Garmestani, A.S., Allen, C.R., Gunderson, L., 2009. Panarchy: discontinuities reveal similarities in the dynamic system structure of ecological and social systems. Ecol. Soc. 14 (1), 15. [online] URL: http://www.ecologyandsociety.org/vol14/iss1/art15/.

Green, O.O., Garmestani, A.S., Allen, C.R., Ruhl, J.B., Arnold, C.A., Gunderson, L.H., Graham, N., Cosens, B., Angeler, D.G., Chaffin, B.C., Holling, C.S., 2015. Barriers and bridges to the integration of social-ecological resilience and law. Front. Ecol. Environ. 13, 332–337.

Gulati, R.D., Pires, L.M.D., Van Donk, E., 2008. Lake restoration studies: failures, bottlenecks and prospects of new ecotechnological measures. Limnologica 38, 233–247.

Gunderson, L.H., 2000. Ecological resilience—in theory and application. Annu. Rev. Ecol. Syst. 31, 425–439.

Gunderson, L.H., Holling, C.S., 2002. Panarchy: Understanding Transformations in Human and Natural Systems. Island Press, Washington, D.C., USA.

Hansson, L.-A., Annadotter, H., Bergman, E., Hamrin, S.F., Jeppesen, E., Kairesalo, T., Luokkanen, E., Nilsson, P.-A., Søndergaard, M., Strand, J., 1998. Biomanipulation as an application of food chain theory: constraints, synthesis, and recommendations for temperate lakes. Ecosystems 1, 558–574.

Holling, C.S., 1973. Resilience and stability of ecological systems. Annu. Rev. Ecol. Syst. 4, 1–23.

Holling, C.S., Allen, C.R., 2002. Adaptive inference for distinguishing credible from incredible patterns in nature. Ecosystems 5, 319–328.

Hooper, D.U., Adair, E.C., Cardinale, B.J., Byrnes, J.E., Hungate, B.A., Matulich, K.L., et al., 2012. A global synthesis reveals biodiversity loss as a major driver of ecosystem change. Nature 486 (7401), 105–108.

Horner-Dixon, T.F., 1991. On the threshold: environmental changes as causes of acute conflict. Int. Secur. 16, 76–116.

Hua, Z.S., Han, Y.J., Chen, L.X., Liu, J., Hu, M., Li, S.J., et al., 2015. Ecological roles of dominant and rare prokaryotes in acid mine drainage revealed by metagenomics and metatranscriptomics. ISME J. 9, 1280.

Hughes, T.P., Carpenter, S., Rockström, J., Scheffer, M., Walker, B., 2013. Multiscale regime shifts and planetary boundaries. Trends Ecol. Evol. 28, 389–395.

Johnson, R.K., Hering, D., 2009. Response of taxonomic groups in streams to gradients in resource and habitat characteristics. J. Appl. Ecol. 46, 175–186.

Johnstone, J.F., Allen, C.D., Franklin, J.F., Frelich, L.E., Harvey, B.J., Higuera, P.E., et al., 2016. Changing disturbance regimes, ecological memory, and forest resilience. Front. Ecol. Environ. 14, 369–378.

Kates, R.W., Clark, W.C., Corell, R., Hall, J.M., Jaeger, C.C., Lowe, I., McCarthy, J.J., Schellnhuber, H.J., et al., 2011. Sustainability Science. Science 292, 641–642.

Klauschies, T., Vasseur, D.A., Gaedke, U., 2016. Trait adaptation promotes species coexistence in diverse predator and prey communities. Ecol. Evol. 6, 4141–4159.

Kühsel, S., Blüthgen, N., 2015. High diversity stabilizes the thermal resilience of pollinator communities in intensively managed grasslands. Nat. Commun. 6, art7989.

Laliberte, E., Wells, J.A., DeClerck, F., Metcalfe, D.J., Catterall, C.P., Queiroz, C., et al., 2010. Land-use intensification reduces functional redundancy and response diversity in plant communities. Ecol. Lett. 13, 76–86.

Lederwasch, A., Mukheibir, P., 2013. The triple bottom line and progress toward ecological sustainable development: Australia's coal mining industry as a case study. Resources 2, 26–38.

Levins, R., Culver, D., 1971. Regional coexistence of species and competition between rare species. Proc. Natl. Acad. Sci. U. S. A. 68, 1246–1248.

Low-Décarie, E., Kolber, M., Homme, P., Lofano, A., Dumbrell, A., Gonzalez, A., Bell, G., 2015. Community rescue in experimental metacommunities. Proc. Natl. Acad. Sci. U. S. A. 112, 14307–14312.

Lyons, K.G., Brigham, C.A., Traut, B.H., Schwartz, M.W., 2005. Rare species and ecosystem functioning. Conserv. Biol. 19, 1019–1024.

Magurran, A.E., 2004. Measuring Biological Diversity. Blackwell, Oxford.

McMichael, A.J., Friel, S., Nyong, A., Corvalan, C., 2008. Global environmental change and health: impacts, inequalities, and the health sector. Br. Med. J. 336, 191–194.

Mitra, C., Kurths, J., Donner, R.V., 2015. A integrative quantifier of multistability in complex systems based on ecological resilience. Sci. Rep. 5, 16196.

Mori, A.S., Furukawa, T., Sasaki, T., 2013. Response diversity determines the resilience of ecosystems to environmental change. Biol. Rev. 88, 349–364.

Mouillot, D., Bellwood, D.R., Barlato, C., et al., 2013. Rare species support vulnerable functions in high-diversity ecosystems. PLoS Biol. 11 (5), e1001569. https://doi.org/10.1371/journal.pbio.1001569.

Mumby, P.J., Chollett, I., Bozec, Y.M., Wolff, N.H., 2014. Ecological resilience, robustness and vulnerability: how do these concepts benefit ecosystem management? Curr. Opin. Environ. Sustain. 7, 22–27.

Nash, K.L., Allen, C.R., Angeler, D.G., Barichievy, C., Eason, T., Garmestani, A.S., Graham, N.A.J., Granholm, D., Knutson, M., Nelson, R.J., Nyström, M., Stow, C.A., Sundstrom, S.M., 2014. Discontinuities, cross-scale patterns, and the organization of ecosystems. Ecology 95, 654–667.

Nash, K.L., Graham, N.A.J., Jennings, S., Wilson, S.K., Bellwood, D.R., 2016. Herbivore cross-scale redundancy supports response diversity and promotes coral reef resilience. J. Appl. Ecol. 53, 646–655.

Nicotra, A.B., Beever, E.A., Robertson, A.L., Hofmann, G.E., O'Leary, J., 2015. Assessing the components of adaptive capacity to improve conservation and management efforts under global change. Conserv. Biol. 29, 1268–1278.

Nyström, M., 2006. Redundancy and response diversity of functional groups: implications for the resilience of coral reefs. Ambio 35, 30–35.

Nyström, M., Folke, C., 2001. Spatial resilience of coral reefs. Ecosystems 4, 406–417.

Oliver, T.H., Heard, M.S., Isaac, N.J.B., Roy, D.B., Procter, D., Eigenbrod, F., Freckleton, R., Hector, A., Orme, C.D.L., Petchey, O.L., et al., 2015. Biodiversity and the resilience of ecosystem services. Trends Ecol. Evol. 30, 673–684.

Palmer, M.A., Menninger, H.L., Bernhardt, E., 2010. River restoration, habitat heterogeneity and biodiversity: a failure of theory or practice? Freshw. Biol. 55, 205–222.

Peterson, G., Allen, C.R., Holling, C.S., 1998. Ecological resilience, biodiversity and scale. Ecosystems 1, 6–18.

Pickett, S.T.A., White, P.S., 1985. The Ecology of Natural Disturbance and Patch Dynamics. Academic Press, New York. White.

Power, D.A., Watson, R.A., Szathmary, E., Mills, R., Powers, S.T., Doncaster, C.P., Czapp, B., 2015. What can ecosystems learn? Expanding evolutionary ecology with learning theory. Biol. Direct 10, 69.

Rist, L., Felton, A., Nyström, M., Troell, M., Sponseller, R.A., Bengtsson, J., Österblom, H., Lindborg, R., Tidåker, P., Angeler, D.G., Milestad, R., Moen, J., 2014. Applying resilience thinking to production systems. Ecosphere 5 (6), art73.

Rockström, J., Steffen, W., Noone, K., Persson, Å., Chapin III, F.S., 2009. Planetary boundaries: exploring the safe operating space for humanity. Ecol. Soc. 14 (2), 32.

Scheffer, M., 1997. Ecology of Shallow Lakes. vol. 22. Springer Science & Business Media.

Scheffer, M., van Nes, E.H., 2007. Shallow lakes theory revisited: various alternative regimes driven by climate, nutrients, depth and lake size. In: Gulati, R.D., Lammens, E., DePauw, N., Van Donk, E. (Eds.), Shallow Lakes in a Changing World. Springer, Dordrecht, pp. 455–466.

Schleuning, M., Fründ, J., García, D., 2015. Predicting ecosystem functions from biodiversity and mutualistic networks: an extension of trait-based concepts to plant–animal interactions. Ecography 38, 380–392.

Smit, B., Wandel, J., 2006. Adaptation, adaptive capacity and vulnerability. Glob. Environ. Chang. 16, 282–292.

Spanbauer, T.L., Allen, C.R., Angeler, D.G., Eason, T., Fritz, S.C., Garmestani, A.S., et al., 2014. Prolonged instability prior to a regime shift. PLoS One 9, e108936.

Spanbauer, T.L., Allen, C.R., Angeler, D.G., Eason, T., Fritz, S.C., Garmestani, A.S., et al., 2016. Body size distributions signal a regime shift in a lake ecosystem. Proc. R. Soc. B 283, 20160249.

Stewart-Koster, B., Olden, J.D., Johnson, P.T.J., 2015. Integrating landscape connectivity and habitat suitability to guide offensive and defensive invasive species management. J. Appl. Ecol. 52, 366–378.

Suding, K.N., Gross, K.L., Houseman, G.R., 2004. Alternative states and positive feedbacks in restoration ecology. Trends Ecol. Evol. 19, 46–53.

Sun, Z.Y., Hai, R., 2011. Ecological memory and its potential applications in ecology: a review. Ying Yong Sheng Tai Xue Bao (Chin. J. Appl. Ecol.) 22 (3), 549–555.

Sundstrom, S.M., Angeler, D.G., Garmestani, A.S., García, J.H., Allen, C.R., 2014. Transdisciplinary application of cross-scale resilience. Sustainability 6 (10), 6925–6948.

Sundstrom, S.M., Eason, T., Nelson, R.J., Angeler, D.G., Allen, C.R., Barichievy, C., Garmestani, A.S., Graham, N.A.J., Granholm, D., Gunderson, L., Knutson, M., Nash, K.L., Nyström, M., Spanbauer, T., Stow, C.A., 2017. Detecting spatial regimes in ecosystems. Ecol. Lett. 20 (1), 19–32.

Thompson, S.K., Seber, G.A.F., 1996. Adaptive Sampling. Wiley, New York.

Uden, D.R., Allen, C.R., Angeler, D.G., Corral, L., Fricke, K.A., 2015. Adaptive invasive species distribution models: a framework for modeling incipient invasions. Biol. Invasions 17, 2831–2850.

Walker, B., Kinzig, A., Langridge, J., 1999. Original articles: plant attribute diversity, resilience, and ecosystem function: the nature and significance of dominant and minor species. Ecosystems 2, 95–113.

Wonkka, C.L., Twidwell, D., West, J.B., Rogers, W.E., 2016. Shrubland resilience varies across soil types: implications for operationalizing resilience in ecological restoration. Ecol. Appl. 26, 128–145.

Further reading

Biggs, R., Carpenter, S.R., Brock, W.A., 2009. Turning back from the brink: detecting an impending regime shift in time to avert it. Proc. Natl. Acad. Sci. U. S. A. 106, 826–831.

Crowl, T.A., Crist, T.O., Parmenter, R.R., Belovsky, G., Lugo, A.E., 2008. The spread of invasive species and infectious disease as drivers of ecosystem change. Front. Ecol. Environ. 6, 238–246.

Eason, T., Garmestani, A.S., Stow, C.A., Rojo, C., Alvarez-Cobelas, M., Cabezas, H., 2016. Managing for resilience: an information theory-based approach to assessing ecosystems. J. Appl. Ecol. 53, 656–665.

Peterson, G.D., 2002. Contagious disturbance, ecological memory, and the emergence of landscape pattern. Ecosystems 5, 329–338.

Resilience Alliance, 2015. Resilience Alliance—Adaptive Capacity. http://www.resalliance. org/adaptive-capacity.

Rodrigues, A.S.L., Brooks, T.M., 2007. Shortcuts for biodiversity conservation planning: the effectiveness of surrogates. Annu. Rev. Ecol. Syst. 38, 713–737.

Scheffer, M., Carpenter, S.R., 2003. Catastrophic regime shifts in ecosystems: linking theory to observation. Trends Ecol. Evol. 18, 648–656.

Spears, B.M., Ives, S.C., Angeler, D.G., Allen, C.R., Birk, S., Carvalho, L., Cavers, S., Daunt, F., Morton, R.D., Pocock, M.J.O., Rhodes, G., Thackeray, S.J., 2015. Effective management of ecological resilience—are we there yet? J. Appl. Ecol. 52, 1311–1315.

Thompson, J.N., Reichman, O.J., Morin, P.J., Polis, G.A., Power, M.E., Sterner, R.W., et al., 2001. Frontiers of ecology. Bioscience 51, 15–24.

Walker, B., Holling, C.S., Carpenter, S.R., Kinzig, A., 2004. Resilience, adaptability and transformability in social–ecological systems. Ecol. Soc. 9 (2), 5.

CHAPTER TWO

Extensive grassland-use sustains high levels of soil biological activity, but does not alleviate detrimental climate change effects

Julia Siebert[a,b,*], Madhav P. Thakur[a,b,c], Thomas Reitz[a,d], Martin Schädler[a,e], Elke Schulz[d], Rui Yin[b,e], Alexandra Weigelt[a,f], Nico Eisenhauer[a,b]

[a]German Centre for Integrative Biodiversity Research (iDiv) Halle-Jena-Leipzig, Leipzig, Germany
[b]Institute of Biology, Leipzig University, Leipzig, Germany
[c]Department of Terrestrial Ecology, Netherlands Institute of Ecology (NIOO-KNAW), Wageningen, The Netherlands
[d]Department of Soil Ecology, Helmholtz Centre for Environmental Research—UFZ, Halle, Germany
[e]Department of Community Ecology, Helmholtz-Centre for Environmental Research—UFZ, Halle, Germany
[f]Department of Systematic Botany and Functional Biodiversity, Institute of Biology, Leipzig University, Leipzig, Germany
*Corresponding author e-mail address: julia.siebert@idiv.de

Contents

Advances in Ecological Research, Volume 60
ISSN 0065-2504
https://doi.org/10.1016/bs.aecr.2019.02.002

25

Abstract

Climate change and intensified land use simultaneously affect the magnitude and resilience of soil-derived ecosystem functions, such as nutrient cycling and decomposition. Thus far, the responses of soil organisms to interacting global change drivers remain poorly explored and our knowledge of below-ground phenology is particularly limited. Previous studies suggest that extensive land-use management has the potential to buffer detrimental climate change impacts, via biodiversity-mediated effects. According to the insurance hypothesis of biodiversity, a higher biodiversity of soil communities and thus an elevated response diversity to climate change would facilitate a more stable provisioning of ecosystem functions under environmental stress. Here we present results of a two-year study investigating, at fine temporal resolution, the effects of predicted climate change scenarios (altered precipitation patterns; passive warming) on three grassland types, differing in land-use intensity, soil biological activity, and in resilience.

We show that future climate conditions consistently reduced soil biological activity, revealing an overall negative effect of predicted climate change. Furthermore, future climate caused earlier and significantly lower peaks of biological activity in the soil. Land-use intensity also significantly decreased soil biological activity, but contrary to general expectations, extensive land use did not alleviate the detrimental effects of simulated climate change. Instead, the greatest reduction in soil biological activity was observed in extensively-used grasslands, highlighting their potential vulnerability to predicted climate change. To assure high levels of biological activity in resilient agroecosystems, extensive land use needs to be complemented by other management approaches, such as the adoption of specific plant species compositions that secure ecosystem functioning in a changing world.

1. Introduction

Climate change is altering the composition of terrestrial ecosystems and the functions they provide (Vitousek, 1994), including soil–driven processes like nutrient cycling and decomposition (Bardgett and van der Putten, 2014; Verhoef and Brussaard, 1990). Climate change not only alters the densities and functional attributes of communities (Blankinship et al., 2011; Briones et al., 2009), but many plant and animal species also adjust their phenology to an extended growing season as a result of a modified climate (Cohen et al., 2018; Menzel and Fabian, 1999; Peñuelas and Filella, 2009). At present, however, we can only speculate on how soil organisms (e.g. microbes and invertebrates) will respond in their year–round activity patterns to changing climate conditions, as we are largely lacking data in high temporal resolution (Bakonyi et al., 2007; Briones et al., 2009;

Eisenhauer et al., 2018; Thakur et al., 2018). Knowledge of belowground activity patterns will be crucial to improve our understanding of key ecosystem functions in a changing world (Bardgett and van der Putten, 2014; Eisenhauer et al., 2018).

Decomposition of soil organic matter involves both soil microbial and invertebrate activity (Swift et al., 1979). While climate change effects on microbial-driven decomposition have been well studied in short-term assessments (A'Bear et al., 2012; Manzoni et al., 2012), the role of soil invertebrates is less well understood (Walter et al., 2013). It is often assumed that decomposition will be enhanced under warmer conditions (Fierer et al., 2005; Rustad et al., 2001), which has found empirical support from lab and field experiments (Conant et al., 2011; Melillo et al., 2002). This is generally in line with predictions of greater metabolic demands of ectothermic organisms at higher temperatures (Gillooly et al., 2001). However, there is mounting evidence that this positive relationship only holds as long as other environmental factors such as soil moisture are not limiting (Butenschoen et al., 2011; Thakur et al., 2018). As soon as higher soil temperatures are accompanied by a decrease in soil moisture, the activity levels of microorganisms and invertebrates decline and thereby potentially slow down decomposition processes (Allison and Treseder, 2008; Davidson and Janssens, 2006; Thakur et al., 2018).

It is evident that drivers of global change do not occur in isolation, but act in concert (Dukes et al., 2005). For instance, changes in temperature and precipitation coincide with dramatic alterations in land use. As the demands for raw materials and food rise with human population growth (Ingram et al., 2008; Tilman et al., 2002), large amounts of land are being converted to arable agriculture and pasture lands subjected to increasingly intensified management (Foley et al., 2005). Such practices include the adoption of a restricted pool of highly productive forage plant species that allow increased mowing frequencies, tillage, and heavy machine use, all of which can impair the functioning of managed ecosystems (Giller et al., 1997; Newbold et al., 2015; Tsiafouli et al., 2015). High tillage and grazing frequencies, and the addition of mineral fertilizers, have been shown to decrease the abundance, diversity, and activity of soil organisms as well as the functions that they drive (Treseder, 2008; Wardle et al., 2002). This means that the strength of climate change effects may depend on the management system, due to the potential interactive effects of climate change and land use (De Vries et al., 2012; Walter et al., 2013).

Extensive management strategies, with less disturbances and greater biodiversity, might be expected to buffer the detrimental effects of climate change and lead to higher resilience of the grasslands, as they represent more complex systems owing to the greater number of species and thus a higher interaction and response diversity (Goldenberg et al., 2018) with an increased likelihood of asynchronous responses of the different species within a trophic group (Craven et al., 2018; Hector et al., 2010; Mazancourt et al., 2013). Intensively-managed, often low-diversity systems, in contrast, are expected to be particularly vulnerable to changing environmental conditions, as predicted by the insurance hypothesis of biodiversity (Isbell et al., 2017; Loreau et al., 2003; Yachi and Loreau, 1999). Indeed, Walter et al. (2013) showed that decomposition rates are more susceptible to drought in grasslands with higher cutting frequency. However, multifactorial studies investigating the interactive effects of climate change and land use on decomposition processes remain scarce (Walter et al., 2013), and we particularly lack insight into the phenological patterns among below-ground soil invertebrates (Eisenhauer et al., 2018). Studying the potential interactive effects of climate change and land-use management across different seasons of the year would help gaining more realistic insights into the temporal dynamics and full-year responses of crucial ecosystem functions in a changing world (Eisenhauer et al., 2018; Bardgett and van der Putten, 2014).

Here, we test for the interactive effects of climate change and land-use management in grasslands on the average levels and phenology (shifts in activity peaks) of soil biological activity and its resilience. We do this by measuring soil microbial respiration (Scheu, 1992) and invertebrate feeding activity (Kratz, 1998), which are tightly linked to decomposition processes in the soil (Thakur et al., 2018), under modified conditions of climate and land-use management. The study was conducted within the framework of the Global Change Experimental Facility (GCEF; Fig. 1A) in Bad Lauchstädt, Germany, a large-scale experimental platform, where predicted climate conditions for the period $\sim$2070–2100 are simulated on 16×24 m-plots: altered temperature (ambient versus ambient $+0.6\,°C$) and precipitation regimes (ambient versus $\sim$20% reduction in summer, $\sim$10% addition in spring and autumn, respectively) are realised with the help of fully automated, fold-out roofs (Schädler et al., 2019). The climate treatments are crossed with three different grassland types in a split-plot design: extensively-used grassland with mowing (mown two times per year, species-rich plant community, no fertilisation, hereafter: extensive meadow), extensively-used grassland with sheep grazing

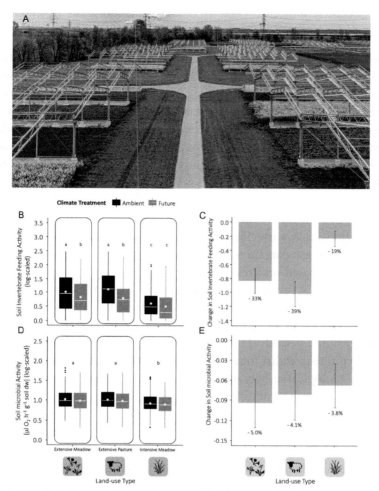

Fig. 1 Interactive effects of climate change and land use on soil biological activity and its resilience. (A) The Global Change Experimental Facility in Bad Lauchstädt, Germany. Image copyright: Tricklabor/Service Drohne. (B) Boxplots showing the interactive effects of climate change and land use on soil invertebrate feeding activity (log-scaled) across all sampling points. (C) Changes in soil invertebrate feeding activity in response to the climate treatment (compared to ambient climate conditions) for the three land-use types. Error bars ± SE based on means (ambient/future) per sampling. (D) Boxplots showing the interactive effects of climate change and land use on soil microbial activity (log-scaled) across all sampling points. (E) Changes in soil microbial activity in response to the climate treatment (compared to ambient climate conditions) for the three land-use types. Error bars ± SE based on means (ambient/future) per sampling. Boxplots show the median (horizontal line), the mean (dot), first and third quartile (rectangle), 1.5 × interquartile range (whiskers), and outliers (isolated points). Letters a, b, and c on top of the boxplots indicate significant differences among treatments based on Tukey's HSD test ($P < 0.05$) run on linear mixed effects models. For interpretation of the references to color, the reader is referred to the online version of this article. $**P < 0.01$, $***P < 0.001$. Black = ambient climate; grey = future climate. Green = extensive meadow (moderately mown); blue = extensive pasture (grazed by sheep); yellow = intensive meadow (frequently mown).

(grazed two to three times per year, species-rich plant community, no fertilisation, hereafter: extensive pasture), and intensively-used grassland (mown three to four times per year, few forage species, mineral fertilizer, hereafter: intensive meadow). All grassland types represent common local management practices including specific species pools and management intervals. As for the land-use types, the climate treatments represent realistic climate scenarios that still allow for inter-annual variability in place of rigid, highly controlled conditions. The experiment started in 2014, and the present study was conducted from March 2015 to April 2017. Measurements were done every three weeks by employing rapid ecosystem function assessment methods, following Thakur et al. (2018), to obtain year-round high temporal resolution data on soil responses. While our current knowledge is predominantly based on a few short-term assessments, this comprehensive study comprises 36 (invertebrate feeding activity) and 34 (microbial activity) sampling dates in two consecutive years to address the need to continuously study the responses of soil organisms to interacting global change drivers over longer time periods (Hamel et al., 2007).

We hypothesised that future climate conditions will change the phenology of soil biological activity by increasing activity in spring and autumn and by reducing activity in summer (Thakur et al., 2018). Thus, we expected a shift towards earlier activity peaks at the beginning of the growing season under future climate conditions. Furthermore, we expected the intensive meadow to show a decrease in soil biological activity, due to frequent disturbances and the use of mineral fertilizer (Treseder, 2008). The most detrimental climate effects were expected for the intensively-managed land-use regime. Accordingly, we expected extensively-used grassland types to show greater resilience by alleviating the detrimental effects of the predicted future climate, due to substantially higher biodiversity (Isbell et al., 2015, 2017), which maintains high levels of soil biological activity.

2. Methods

2.1 Study design

The Global Change Experimental Facility was established in 2013 to study the interactive effects of climate change (including elevated temperature and changes in precipitation patterns) and land-use intensity on managed terrestrial ecosystems using realistic scenarios (Schädler et al., 2019). The study site is located at the field research station of the Helmholtz-Centre for Environmental Research (UFZ) in Bad Lauchstädt, Germany

(51° 22′ 60 N, 11° 50′ 60E, 118 m a.s.l.), and was formerly used as an arable field (last crop cultivation in 2012). Being located in the Central German dry area (Querfurter Platte), the site has a mean annual precipitation of 489 mm (1896–2013) and a mean annual temperature of 8.9 °C (1896–2013). The soil is a Haplic Chernozem with a humus layer reaching down to more than 40 cm depth. This highly fertile soil type was developed upon carbonatic loess substrates (around 70% silt and 20% clay content). The soil is known for its high water-retention capacity (nearly reaching the mean annual precipitation), ensuring comparatively low susceptibility to drought stress (Altermann et al., 2005). Within the upper 15 cm, pH values ranged from 5.8 to 7.5, while total carbon and total nitrogen varied between 1.71 and 2.09% and 0.15–0.18%, respectively.

The experiment consisted of 50 plots arranged in 10 mainplots (Schädler et al., 2019). The two experimental treatments were implemented in a split-plot design with the climate treatment carried out on the mainplot level ($n = 10$) and the land-use treatment implemented on the plot level ($n = 50$), randomly arranged within the mainplots. Thus, for each of the five land-use types, there are five plots with future climate conditions and five plots with ambient climate conditions that serve as a climate control. The spatial scale is realised by a large plot size of 16×24 m that allows the use of standard agricultural equipment. Each plot has a buffer zone of 2 m to the eastern and western sides and 4.5 m to the southern and northern sides. In addition to the buffer zones, the randomised location of plots within mainplots mimics all possible neighbourhood arrangements of land-use treatments (Schädler et al., 2019). All measurements took place in the inner plot area (15×12 m) on a specific transect for soil measurements (Fig. 2).

The climate treatments were first applied in 2014 (spring 2014: start of temperature treatment; summer 2014: start of precipitation treatment). All mainplots were equipped with a steel framework of 5 m height that allowed the mounting of equipment to impose the climate treatment. In the case of the control mainplots, the steel framework served as a control for potential infrastructure effects, such as microclimatic effects. In the case of the mainplots assigned to the climate treatment, the roof constructions included an irrigation system and mobile roof and side panels that can be closed via rain sensors/timers.

The climate treatment was chosen based on a consensus scenario across several dynamic models for Central Germany for 2070–2100, which include higher inter-annual rainfall variability with longer drought periods over summer and increased precipitation in spring and autumn (Doscher et al., 2002;

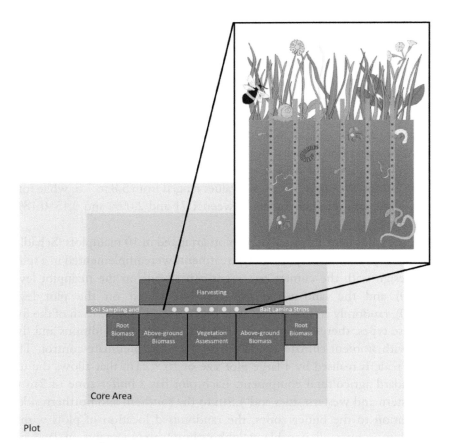

Fig. 2 The plot design of the Global Change Experimental Facility (GCEF) in Bad Lauchstädt, Germany. All 50 plots of the GCEF (24 × 16 m) have an inner core area (15 × 12 m) where the samplings took place. The area for soil measurements describes a north-south transect (in blue, 15 × 0.5 m in total). Six bait lamina strips were placed on the transect in a distance of approximately 20 cm to each other. Above-ground biomass was harvested on a 9 × 1.6 m area. Root biomass was sampled on 2 m × 2 m areas using multiple soil cores.

Jacob and Podzun, 1997; Rockel et al., 2008). Furthermore, the mean annual temperature is predicted to rise by up to 3 °C (Kerr, 2007; Meinke et al., 2010). The treatments are applied as realistic climate scenarios, which include natural variability instead of highly controlled conditions. Thus, the climate treatments are applied relative to the ambient conditions and allow for inter-annual variability. Over summer (June–August), the climate treatment includes a reduction of precipitation by ∼20% (closing of roof and side panels via rain sensors). In spring (March–May) and autumn

(September–November), the precipitation is increased by ~10% by the use of an irrigation system that uses water from a large rain water reservoir (Fig. 3).

Passive warming by reducing heat emission during the night with mobile roofs became a standard method in climate change experiments and mimics

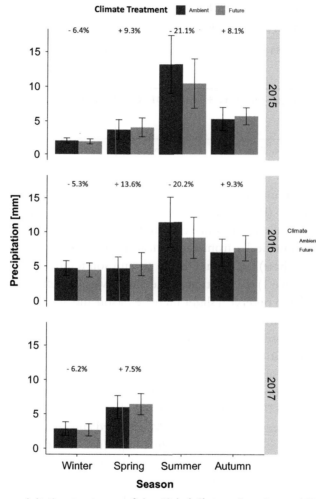

Fig. 3 The precipitation treatment of the Global Change Experimental Facility. Panels show ambient and future climate conditions for each season and year in which the study took place. Intended reduction in summer: ~20%, intended increase in spring and autumn: ~10% (small reduction in winter can be explained by a decrease of vegetation cover on plots with future climate, see Schädler et al. (2019)). Dark grey = ambient climate; light grey = future climate. Percentages above the bar plots indicate the changes in precipitation under future climate conditions compared to ambient climate conditions. Error bars ±SE.

Table 1 The temperature treatment of the global change experimental facility. Shown are average deviations of daily means in air and soil temperature for future climate conditions compared to ambient climate conditions during active and inactive roof phases within the study period ($\pm$ SE). Values different from zero (t-test, $P < 0.05$) are given in bold.

	June 1st 2015[a]— December 11th 2015	December 12th 2015—March 23rd 2016	March 24th 2016—November 22nd 2016	November 23rd 2016—December 31st 2017
	Roofs active	Roofs inactive	Roofs active	Roofs inactive
Daily mean air temperature at 5 cm height	**+0.55 ± 0.03 °C**	+0.006 ± 0.006 °C	**+0.55 ± 0.03 °C**	**+0.08 ± 0.03 °C**
Daily mean soil temperature at 1 cm depth	**+0.65 ± 0.02 °C**	**+0.11 ± 0.01 °C**	**+0.44 ± 0.02 °C**	**+0.31 ± 0.03 °C**

[a]Active roof phase started on February 25th, but data from temperature loggers is only available from June 2015 onwards.

the realistic phenomenon of stronger increases of minimal rather than maximal temperatures (Beier et al., 2004). In our experiment, passive warming overnight was realised on future climate mainplots by closing the roof and side panels automatically from calendrical sundown to sunrise (within the active roofing phase), which increased air and soil temperature on average by ~0.6 °C (Table 1). In addition to the direct warming effects of roof closing, we could also observe a slight increase of soil and air temperatures near the soil surface during inactive roofing phases. This can be explained by a decrease of vegetation cover on plots with future climate and the accompanying direct insolation on the soil surface (Schädler et al., 2019). This can be interpreted as a realistic side effect of climate change on the microclimate.

In 2015, the roofs were in place from February 25th to December 11th (roofs were closed 79% of night time, deviations were due to frost and wind). In 2016, the roofing phase started on March 23rd and ended on November 23rd (roofs were closed 82% of night time). In 2017, the roofing phase started on March 11th. The exact timing of the inactive roofing phase over winter (i.e. no climate treatment intended) was determined based on the forecasts of longer frost periods. Within the active roofing phase, roofs and side panels open automatically at wind speeds above $7 \, \text{ms}^{-1}$ and frost to avoid damages (Schädler et al., 2019).

Within each mainplot there were plots representing five different land-use treatments, each with different levels of land-use intensity: (1) conventional farmland; (2) organic farmland; (3) intensive meadow; (4) extensive meadow; and, (5) extensive pasture (the two last-mentioned summarised as extensively-used grasslands). In place of an experimental gradient, each land-use scenario represented a commonly-used management type in the locality around Bad Lauchstädt that has specific combinations of plant diversity and community structure, management intervals, and fertilisation.

The intensive meadow consisted of typical forage plant species, fertilised with mineral fertilizer and mown frequently (three times in 2015, four times in 2016). The extensively-used grasslands were either mown at a moderate frequency (two times in 2015 and 2016, no fertilisation) or moderately grazed by sheep (two times in 2015, three times in 2016, by a group of 20 sheep grazing on each plot for 24 h). After each mowing event, the cut plant biomass was removed from the plots as would normally happen during hay harvest. In the extensive meadow, the mown biomass was left on the plots for some days to enable natural shed of seeds back onto the soil. While the extensively-used grasslands contained 53.1 ± 2.0 plant species per $9 \, m^2$, the intensive meadow had 10.1 ± 3.6 plant species per $9 \, m^2$ (mean $\pm$ SD for 2015–2017). For details on the sown plant species pools, see Schädler et al. (2019). Owing to methodological constraints in making the soil measurements, due to management such as repeated ploughing events, and to a focus on the hypothesis for the potential buffering effects of more species-rich, extensively-used grasslands compared to intensively-used grasslands (De Vries et al., 2012; Isbell et al., 2015), the data presented in this paper is restricted to the extensive meadows, the extensive pastures, and the intensive meadows (i.e. 30 plots arranged within the 10 mainplots) and does not include the two farmlands.

2.2 Soil animal feeding activity and soil microbial activity

We monitored soil invertebrate feeding activity and soil microbial activity every three weeks over a two-year period (March 2015 to April 2017), resulting in 36 and 34 sampling time points with 1080 and 1020 observations of invertebrate feeding activity and microbial activity, respectively. Soil invertebrate feeding activity was assessed using the bait lamina test (Terra Protecta GmbH, Berlin, Germany) as a commonly-used rapid ecosystem function assessment method (Kratz, 1998; Thakur et al., 2018). The bait strips are made of PVC ($1 \, mm \times 6 \, mm \times 120 \, mm$) and have 16 holes (1.5 mm diameter). Original sticks were ordered from Terra Protecta and filled with an artificial organic bait substrate, which was prepared according to the recommendations of Terra Protecta, consisting of 70% cellulose powder, 27% wheat bran, and 3% activated carbon. The bait substrate is primarily consumed by mites, collembolans, enchytraeids, millipedes, and earthworms

(Gardi et al., 2009; Hamel et al., 2007; Harding and Stuttard, 1974), whereas microbial activity plays a minor role in bait substrate loss (Hamel et al., 2007; Rożen et al., 2010; Simpson et al., 2012).

The bait lamina strips were inserted vertically into the soil with the uppermost hole just beneath the soil surface. A steel knife was used to make a slot in the soil into which the strips were carefully inserted. Six strips were placed at a distance of approximately 20 cm from one another per plot to account for spatial heterogeneity (Fig. 2). After three weeks of exposure, the bait lamina strips were removed from the soil, directly evaluated in the field, and replaced by a new bait strip. Each hole was carefully inspected and rated as 0 (no invertebrate feeding activity), 0.5 (intermediate feeding activity), or 1 (high invertebrate feeding activity). Soil invertebrate feeding activity can therefore range from 0 (no feeding activity) to 16 (maximum feeding activity) per strip. Mean bait consumption of the six strips was calculated per plot prior to statistical analysis.

To measure soil microbial activity, soil samples were taken every three weeks using a steel corer (1 cm diameter; 15 cm deep). Seven subsamples per plot were homogenised, sieved through a 2 mm sieve, and stored at 4 °C. Basal respiration (without addition of substrate) was measured using an O_2-microcompensation apparatus in the lab (Scheu, 1992). Soil microbial respiration was measured at hourly intervals for 24 h at 20 °C (i.e. at constant temperature), and basal respiration as a measure of microbial activity was calculated as the mean O_2 consumption rate 14–24 h after the start of the measurements ($\mu l\ O_2\ h^{-1}$ per g soil dry weight). Since we were able to monitor the phenology of invertebrate feeding activity in situ, we focused on these results hereafter. Nevertheless, we performed time series analyses for both response variables (Table 2).

Table 2 Results from generalised additive mixed-effects models (GAMMs) for soil invertebrate feeding activity and soil microbial activity (both log-scaled). Climate, land use, and time were incorporated as smooth terms. Statistically significant results are given in bold.

Treatment	Soil invertebrate feeding activity			Soil microbial activity		
	edf	F-value	P-value	edf	F-value	P-value
s (Time, Land use)	2.63	1.38	0.24	17.61	0.47	0.98
s (Time, Climate)	**21.91**	**4.55**	**<0.0001**	12.83	0.08	1
s (Time, Land use, Climate)	**24.63**	**6.74**	**<0.0001**	**3.96**	**6.92**	**<0.0001**

2.3 Assessments of potential explanatory variables

On the two meadow types, plant shoot biomass was assessed on subplots of 9 m × 1.6 m (Fig. 2) using a rotary mower with a cutting height of 5–8 cm. The intensive meadow was mown four times per year, whereas the extensive meadow was mown twice in 2016. On the extensive pasture (i.e. sheep grazing), such data is not available. Plant root biomass was sampled in April and June 2016 on all three grassland types. Using a soil corer of 3.5 cm diameter, four subsamples were taken per plot (see Fig. 2 for the specific location) at a depth of 0–15 cm. All subsamples were pooled and repeatedly rinsed in water to obtain the fraction of fine roots (<2 mm), which was then dried at 70 °C. Root mass density was calculated per dm^3 soil.

To analyse soil microbial biomass and abiotic soil parameters, soil samples were taken in April, June, and October 2016 by using a steel corer (1 cm diameter; 15 cm deep). Seven subsamples per plot were homogenised, sieved at 2 mm, and stored at 4 °C. An O_2-microcompensation system (Scheu, 1992) was used to estimate the maximal respiratory response of soil microorganisms following the addition of a glucose standard (4 mg g^{-1} dry weight soil, solved in 1.5 mL distilled water) to determine soil microbial biomass (μg Cmic g^{-1} dry weight soil). Gravimetric soil moisture contents were determined using a fully automatic moisture analyser (Kern DBS60–3 from Kern & Sohn GmbH, Germany). Soil pH was measured with a pH electrode (Mettler Toledo InLab Expert Pro-ISM) after shaking the soil for 1 h in 0.01 M $CaCl_2$ (1:2.5 w/v). Hot water extractable carbon (HWC) and nitrogen (HWN), which represent the labile organic C and N pools, were determined from 10 g of air-dried soil following the method of Schulz (2002) and analysed using an elemental analyser for liquid samples (Multi N/C, Analytik Jena, Germany).

Soil mesofauna (mostly Collembola and Acari) was sampled in June and October 2016. Three soil cores (6 cm diameter, 5 cm depth) were randomly taken per subplot on a strip of 15 m × 0.5 m (Fig. 2) and extracted in a MacFadyen high-gradient extractor for 10 days (Macfadyen, 1961) before the abundances were determined.

2.4 Statistical analysis

Soil invertebrate feeding activity was measured at 36 time points, and soil microbial activity was measured at 34 time points. Given that we expected treatment effects to vary with time in a nonlinear way, we used generalised additive mixed-effects models (GAMMs) to test the interactive effects of climate, land use, and time on soil biological activity. We chose GAMMs due to their flexibility in including smooth functions of covariates without restricting the relationships to be linear, quadratic, or cubic. The model structure of the GAMM was: soil biological activity ~s (time, climate) * s (time, land use) * s (time, climate, land use) + (1 | mainplot/plot), with 's' indicating smoothing functions for GAMM. Experimental plots were nested within mainplots and incorporated as a random intercept for the experimental design. We applied GAMMs using the 'gamm4' package (Wood and Scheipl, 2017). The test statistics for GAMMs were obtained from the 'itsadug' package (van Rij et al., 2017). In addition, we compared the results from GAMMs with linear mixed effects models (LMMs), in which we tested the treatment effects across time on soil biological activity using the package 'nlme' (Pinheiro et al., 2017). For LMMs, a random intercept with mainplots nested within sampling time points, nested within years was included in the models. We accounted for repeated measurements by including a compound symmetry covariance structure, which fitted the data better than a first-order autoregressive covariance structure based on the difference in their Akaike information criterion (AIC) value. Invertebrate feeding activity and microbial activity were log-transformed ($\log (x + 1)$) to improve the fit of the model. The raw means and standard errors of both response variables are presented in Table 3.

The 'quantmod' package (Ryan et al., 2017) was used to identify the nearest peaks (using the 15 closest data points) in the time series data of soil invertebrate feeding activity under both ambient and future climate conditions.

Table 3 Mean values of soil invertebrate feeding activity and soil microbial activity. Shown are means of the non-transformed data ($\pm$ SE) for the different land-use types (extensive meadow, extensive pasture, and intensive meadow) under ambient and future climate conditions.

	Extensive meadow				Extensive pasture				Intensive meadow			
	Ambient		Future		Ambient		Future		Ambient		Future	
	mean	$\pm$SE	mean	$\pm$SE	mean	$\pm$SE	mean	$\pm$SE	mean	$\pm$SE	mean	$\pm$SE
Soil invertebrate feeding activity	2.51	0.20	1.69	0.13	2.61	0.16	1.59	0.13	1.10	0.10	0.89	0.09
Soil microbial activity	1.87	0.06	1.78	0.06	1.84	0.06	1.77	0.06	1.59	0.05	1.53	0.05

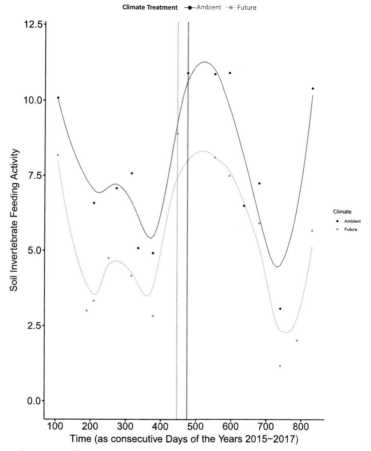

Fig. 4 Peak values of soil invertebrate feeding activity under ambient (dark grey) and future (light grey) climatic conditions within the entire measurement period. The nearest peaks were identified using the 15 closest data points. The grey shaded area represents the difference in days between the two highest peak values of invertebrate feeding activity for the two climate scenarios (29 days).

After the identification of all peaks in the time series (i.e. the local maxima after restricting the search to the nearest 15 data points for each climate condition), we calculated the day difference between the two highest peaks (among all the identified peaks, see Fig. 4) of invertebrate feeding activity for the two climate scenarios. Finally, LMMs (Pinheiro et al., 2017) were used to analyse the effects of climate, land use, season, and their interactions on potential explanatory

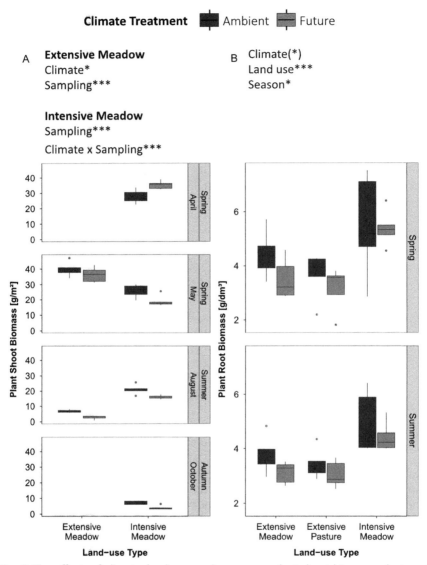

Fig. 5 The effects of climate, land use, and season on plant shoot biomass, plant root biomass, soil microbial biomass, and soil water content. (A) Plant shoot biomass. Extensive meadow was mown two times (May and August); intensive meadow was mown four times (April, May, August, and October). Both land-use types were analysed separately. (B) Plant root biomass from 0 to 15 cm depth. Spring = April; summer = June. (C) Soil microbial biomass. Measured using substrate-induced respiration (Scheu, 1992). Spring = April; summer 1 = June; autumn = October. (D) Soil water content. Spring = April; summer = June; autumn = October. Data from 2016 were used. EM = extensive meadow; ES = extensive pasture; IG = intensive meadow. Dark grey = ambient climate; light grey = future climate. (*) $P < 0.1$; *$P < 0.05$; ***$P < 0.001$.

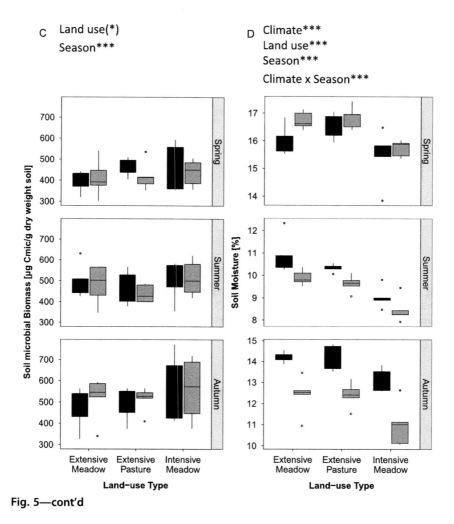

Fig. 5—cont'd

variables (i.e. plant root biomass, microbial biomass, soil water content, soil fauna groups, pH, available C and N) using data available for spring, summer, and autumn 2016 (Figs 5–7). In the case of plant shoot biomass, the effects of climate and sampling were analysed separately for each land-use type because of the different mowing frequencies. A random intercept with plots nested within mainplots was included in the models. All statistical analyses were performed using the R statistical software version 3.5.1 (R Core Team, 2017).

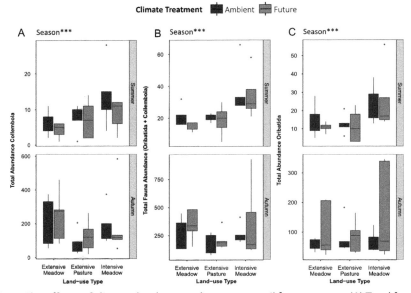

Fig. 6 The effects of climate, land use, and season on soil fauna groups. (A) Total fauna (Oribatida+Collembola). (B) Collembola. (C) Oribatida. Data from 2016 were used: summer = June; autumn = October. EM = extensive meadow; ES = extensive pasture; IG = intensive meadow. Dark grey = ambient climate; light grey = future climate. ***P < 0.001.

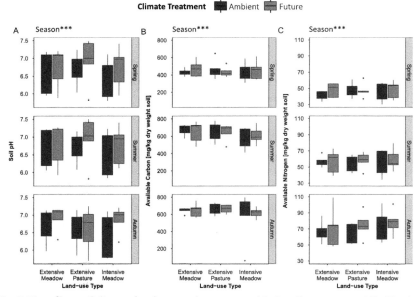

Fig. 7 The effects of climate, land use, and season on abiotic soil parameters. (A) pH-value. (B) Available carbon (hot-water extractable fraction). (C) Available nitrogen (hot water extractable fraction). Data from 2016 were used: spring = April; summer = June; autumn = October. EM = extensive meadow; ES = extensive pasture; IG = intensive meadow. Dark grey = ambient climate; light grey = future climate. ***P < 0.001.

3. Results

3.1 Land use and climate change effects on soil biological activity

Overall, future climate conditions significantly reduced soil invertebrate feeding activity in both extensively-used grasslands, whereas the intensive meadow showed the lowest activity levels under both climate conditions without significant differences (Fig. 1B; Table 4). The strongest climate change-induced reduction was observed in the extensive pasture, the weakest reduction in the intensive meadow (Fig. 1C). Similarly, soil microbial activity was significantly reduced by future climate conditions across all grassland types, with significantly lower activity levels in intensive meadows than in the two extensively-used grasslands (Fig. 1D; Table 4). The strongest climate change-induced reduction was observed for the extensive meadow, the weakest reduction in the intensive meadow (Fig. 1E).

3.2 Interactive effects of climate, land use, and time on soil biological activity

We found a significant three-way interaction effect of time, climate, and land use on both soil invertebrate feeding activity and soil microbial activity (GAMM, Table 2). Future climate conditions decreased soil invertebrate feeding activity for most parts of the study period, except for spring 2016 (Fig. 8). However, changes of average levels of activity depended on the season though. Soil invertebrate feeding activity showed an earlier, but lower peak under future climate conditions in spring 2016. That is, the overall

Table 4 Results of linear mixed effects models (LMMs) for the effects of climate, land use, and their interaction on soil invertebrate feeding activity and soil microbial activity (both log-scaled) across all samplings. Mainplots nested within samplings nested within years served as a random intercept in the model. A compound symmetry covariance structure was used to account for repeated measurements. F-values are given with numerator and denominator degrees of freedom. Significant results are shown in bold.

	Climate		Land use		Climate x Land use	
	F-value	P-value	F-value	P-value	F-value	P-value
Soil invertebrate feeding activity	50.61 (1, 323)	**<0.0001**	102.51 (2, 712)	**<0.0001**	6.47 (2, 712)	**0.0016**
Soil microbial activity	16.16 (1, 305)	**0.0001**	65.30 (2, 666)	**<0.0001**	0.11 (2, 666)	0.90

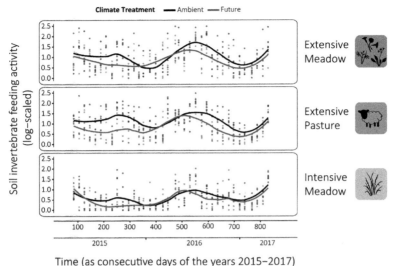

Time (as consecutive days of the years 2015–2017)

Fig. 8 Temporal patterns of the interactive effects of climate and land use on soil invertebrate feeding activity. The panels show the treatment effects over time, separated by the three land-use types (climate × land use × time: $F = 6.74$, $P \leq 0.0001$). The curves are based on the 'loess' smoothing function from the 'ggplot2' package in R with lambda = 0.4. For interpretation of the references to color, the reader is referred to the online version of this article. Black = ambient climate; grey = future climate. Green = extensive meadow (moderately mown); blue = extensive pasture (grazed by sheep); yellow = intensive meadow (frequently mown).

highest value of invertebrate feeding activity under future climate conditions was reached 29 days earlier than the overall highest value recorded under ambient climate conditions (Figs 4 and 8). Despite seasonal fluctuations, intensive meadows showed the lowest activity patterns under both climate conditions throughout the study period. With respect to the interactive effects of land use and climate, intensive meadows were less affected by climate change compared to the two extensively-used grassland types, which showed an equally strong decline (Fig. 8). Soil microbial activity showed similar, although weaker responses to the treatments (Fig. 9).

3.3 Responses of additional plant and soil variables

Future climate consistently reduced plant shoot biomass on the extensive meadow. The intensive meadow benefitted from future climate conditions in April, but shoot biomass was strongly reduced by the climate treatment on all other harvest dates in 2016 (Fig. 5A; Table 5). Plant root biomass was strongly affected by land use, showing the highest biomass under intensive

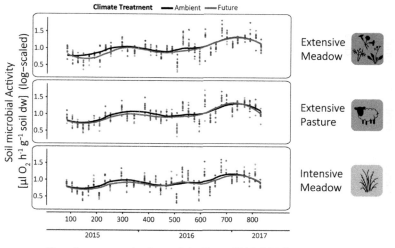

Fig. 9 Temporal patterns of the interactive effects of climate, land use, and time on soil microbial activity. The panels show the treatment effects over time, separated by the three land-use types (climate × land use × time: $F = 6.92$, $P \leq 0.0001$). The curves are based on the 'loess' smoothing function from the 'ggplot2' package in R with lambda = 0.4. For interpretation of the references to color, the reader is referred to the online version of this article. Black = ambient climate; grey = future climate. Green = extensive meadow (moderately mown); blue = extensive pasture (grazed by sheep); yellow = intensive meadow (frequently mown).

Table 5 Results of linear mixed effects models (LMMs) for the effects of climate, sampling, and their interaction on plant shoot biomass. Plots nested within mainplots served as a random intercept in the model. Each land-use type was analysed separately because of the different mowing frequencies (Extensive meadow: May and August 2016; Intensive meadow: April, May, August, October 2016). F-values are given with numerator and denominator degrees of freedom. Significant results are shown in bold.

	Climate		Sampling		Climate × Sampling	
	F-value	**P-value**	**F-value**	**P-value**	**F-value**	**P-value**
Extensive meadow	6.25 (1, 8)	**0.04**	476.39 (1, 8)	**<0.0001**	0.02 (1, 8)	0.89
Intensive meadow	2.42 (1, 8)	0.16	146.05 (3, 24)	**<0.0001**	13.23 (3, 24)	**<0.0001**

management. There was some evidence of an effect, at the 10% level of significance, of future climate on plant root biomass, especially leading to reduced biomass on the extensively-used grasslands (Fig. 5B; Table 6). Another trend was observed for land-use effects on soil microbial biomass,

Table 6 Results of linear mixed effects models (LMMs) for the effects of climate, land use, season, and their interactions on additional explanatory variables. Data from spring, summer, and autumn 2016 were included. Plots nested within mainplots served as a random intercept in the model. F-values are given with numerator and denominator degrees of freedom. Significant results are shown in bold.

	Climate		Land use		Season		Climate x Land use		Climate x Season		Land use x Season		Climate x Land use x Season	
	F-value	P-value	F-value	P-value	F-value	P-value	F-value	P-value	F-value	P-value	F-value	P-value	F-value	P-value
Plant root biomass [g/dm³]	4.16 (1, 8)	0.08	18.04 (2, 16)	**0.0001**	5.23 (1, 24)	**0.03**	0.32 (2, 16)	0.73	0.0001 (1, 24)	0.99	0.70 (2, 24)	0.51	0.24 (2, 24)	0.79
Microbial biomass [µg Cmic/g dry weight soil]	0.0041 (1, 8)	0.95	3.13 (2, 16)	0.07	19.94 (2, 48)	**<0.0001**	0.42 (2, 16)	0.66	0.75 (2, 48)	0.48	1.76 (4, 48)	0.15	0.82 (4, 48)	0.52
Total abundance soil fauna	1.38 (1, 16)	0.26	1.88 (2, 16)	0.19	57.55 (1, 24)	**<0.0001**	0.06 (2, 16)	0.94	1.71 (1, 24)	0.20	1.45 (2, 24)	0.26	0.07 (2, 24)	0.93
Total abundance Collembola	0.27 (1, 16)	0.61	2.00 (2, 16)	0.17	42.46 (1, 24)	**<0.0001**	0.06 (2, 16)	0.94	0.40 (1, 24)	0.54	2.06 (2, 24)	0.15	0.04 (2, 24)	0.96
Total abundance Oribatida	2.47 (1, 16)	0.14	0.96 (2, 16)	0.40	23.59 (1, 24)	**0.0001**	0.73 (2, 16)	0.50	2.69 (1, 24)	0.11	0.21 (2, 24)	0.82	0.64 (2, 24)	0.54
Soil moisture [%]	33.53 (1, 8)	**0.0004**	41.96 (2, 16)	**<0.0001**	904.38 (2, 48)	**<0.0001**	0.05 (2, 16)	0.95	29.79 (2, 48)	**<0.0001**	1.48 (4, 48)	0.22	0.43 (4, 48)	0.79
Soil pH	0.40 (1, 8)	0.55	1.62 (2, 16)	0.23	16.69 (2, 43)	**<0.0001**	0.09 (2, 16)	0.91	0.93 (2, 43)	0.40	1.33 (4, 43)	0.28	0.16 (4, 43)	0.96
Available carbon [mg/kg dry weight soil]	0.03 (1, 8)	0.86	2.58 (2, 16)	0.11	69.66 (2, 48)	**<0.0001**	0.59 (2, 16)	0.57	0.07 (2, 48)	0.93	0.78 (4, 48)	0.54	0.30 (4, 48)	0.88
Available nitrogen [mg/kg dry weight soil]	0.27 (1, 8)	0.62	0.97 (2, 16)	0.40	129.87 (2, 48)	**<0.0001**	0.03 (2, 16)	0.97	0.83 (2, 48)	0.44	1.33 (4, 48)	0.27	0.87 (4, 48)	0.49

resulting in highest levels under intensive management, and soil microbial biomass varied significantly among seasons, with highest values in autumn and lowest values in spring (Fig. 5C; Table 6). Soil water content was higher on future climate plots in spring, but decreased substantially in summer and autumn. Furthermore, the intensive meadow had the lowest soil water content of all land-use types (Fig. 5D; Table 6). The densities of detritivores (collembolans and oribatid mites; Fig. 6, Table 6) as well as pH-value, available carbon, and available nitrogen (Fig. 7; Table 6) varied significantly among seasons, but were not significantly affected by the experimental treatments in 2016.

4. Discussion

The present study reveals that two key response variables for soil biological activity showed similar responses to climate change and land-use management, which are the two main global change drivers affecting terrestrial ecosystems (Sala et al., 2000). By covering a full two-year period, with continuous measurements across all seasons, we show that future climate conditions and intensive land use can be expected to significantly reduce soil biological activity with changing magnitudes across time. Extensive-land use sustained high levels of soil biological activity, but contrary to our expectations and general assumptions, did not alleviate the detrimental effects of climate change. Instead, the decrease in soil biological activity under future climate conditions was most pronounced under extensive management.

Overall, our results show that climate change, simulated by a combination of $\sim 0.6\,^\circ C$ warming and shifts in precipitation patterns across the growing season, consistently reduced soil biological activity throughout the study period, while the magnitude of this reduction depended on the season. This was likely caused by a significantly lower soil water content under the future climate scenario (Fig. 5D), in line with the general expectation that a concurrence of warming and reduced summer rainfall has detrimental effects on soil biological activity (Allison and Treseder, 2008; Davidson and Janssens, 2006; Thakur et al., 2018). We found that the detrimental effects of summer drought exceeded those of elevated precipitation in spring and autumn.

There are potentially a number of explanations for the lower level of invertebrate feeding activity in response to the climate treatment. First, many soil organisms live in a pore system with extensive surfaces of water films and are depending on a water-saturated atmosphere, which might be scarce under future climate conditions (Coleman et al., 2004; Verhoef and Brussaard, 1990).

Second, mobile soil organisms may move to more favourable habitats, e.g., to deeper soil layers (Briones et al., 2007), therefore not contributing to bait perforation within the upper 15 cm that we sampled with our methods. Third, there might be indirect climate effects on soil organisms via altered substrate characteristics, as dry soil is more difficult to ingest and to digest for soil invertebrates (Thakur et al., 2018). The latter explanation may be further exacerbated by the increased metabolic demands of soil organisms under warmed conditions (Brown et al., 2004), which put additional pressure on foraging success under already detrimental conditions (Thakur et al., 2018).

Surprisingly, we did not find higher soil biological activity levels in spring and autumn under future climate conditions, which were expected based on increased precipitation during these seasons. Substantially higher percentages of bare soil cover on future climate plots (Fig. 10) and lower plant shoot and root biomass (Fig. 5A and B) may explain these results. Bare soil is known to strongly reduce bacterial and fungal decomposition rates as well as soil enzyme activities, which may lead to bottom-up-induced changes at higher trophic levels, highlighting the importance of vegetation cover for soil biological activity (Birkhofer et al., 2011; Moreno et al., 2009; Sánchez-Moreno et al., 2015). Summer drought-induced reductions in vegetation cover are unlikely to be compensated by elevated growth in other seasons (Fig. 5A); thus, soil biological activity on future climate plots may still be restrained during more favourable (wet) conditions in spring and autumn. This finding suggests that shifts in precipitation patterns can impair crucial ecosystem functions without any significant net changes in annual precipitation amounts.

In March 2016, we detected a shift towards earlier, and lower peaks in invertebrate feeding activity under future climate conditions (29 days; Figs 4 and 8), indicating a potential shift in the phenology of soil biological activity. This phenological shift may be a direct response to altered soil moisture and temperature and/or indirectly mediated via changes in plant phenology in response to the climate treatment (Eisenhauer et al., 2018). As the timing of plant inputs to the soil advances earlier in the growing season under climate change (Delbart et al., 2008; Menzel and Fabian, 1999; Nord and Lynch, 2009), decomposers are forced to synchronize with plants to optimize resource consumption. This is also supported by higher plant shoot biomass on intensive meadows under future climate in spring 2016 (Fig. 5A), which may be a consequence of accelerated plant growth early in the year. Shifts in soil biological activity may also induce alterations in nutrient cycling by changing the time at which nutrients are made available

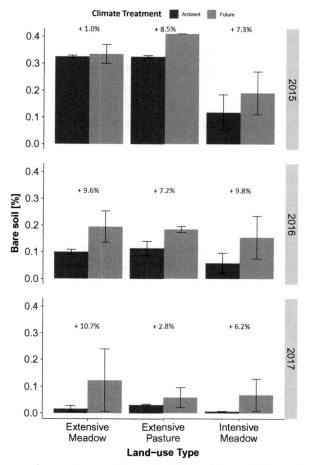

Fig. 10 Percentage bare soil cover. All panels show ambient and future climatic conditions for each year in which the study took place. Dark grey = ambient climate; light grey = future climate. Percentages above the bar plots indicate the increase in bare soil under future climatic conditions compared to ambient climatic conditions. Error bars ±SE.

for plants (Bardgett and Wardle, 2010). Where certain groups of organisms follow this climate-induced shift in phenology and others do not, this effect could potentially lead to temporal mismatches between above- and below-ground components of the community that affect ecosystem functioning (Eisenhauer et al., 2018) and may even lead to changes in community composition and species extinctions (Durant et al., 2007; Thackeray et al., 2016). Temporal mismatches might alter interactions among trophic groups, such as predator-prey relationships, whose synchrony is crucial for natural pest control in agricultural systems (Durant et al., 2007; Thomson et al., 2010). We did, however, not detect any significant changes in detritivore densities in response to the climate and land-use treatments, which may have been due to the snap-shot nature of these assessments. Nevertheless, our study provides compelling evidence for climate change-induced shifts in the phenology of soil biological activity, mainly for soil invertebrate activity patterns. We encourage future studies to investigate the causes and consequences of phenological shifts below-ground for terrestrial ecosystem functioning, which has rarely been considered compared to phenological shifts in above-ground taxa (Eisenhauer et al., 2018).

Extensive management is known to facilitate the abundance and diversity of a wide range of organisms above and below the ground (Bengtsson et al., 2005). This was for example shown for soil organisms, as enhanced microbial activity and diversity of arbuscular-mycorrhizal fungi (França et al., 2007) or higher densities of invertebrate predators (Birkhofer et al., 2008) and carabid beetles (Döring and Kromp, 2003). In the case of grasslands, these positive effects can be connected to higher plant species richness, typically realised in conditions of extensive management that promote soil organisms and processes (Balvanera et al., 2006; Eisenhauer et al., 2013; Lange et al., 2015). Land-use intensification, on the other hand, has been found to have negative effects on soil fauna (Birkhofer et al., 2012; Decaëns and Jiménez, 2002) by reducing taxonomic richness, subsequently leading to less complex and functionally depauperate soil communities (Tsiafouli et al., 2015). Indeed, our results confirm that intensive land use led to the lowest activity levels throughout the study period. Our study also shows that extensive management supported higher absolute levels of soil biological activity than intensive management under both climate scenarios, but with no effect of extensive mowing and sheep grazing. This would support the value of extensive grassland management for enhancing soil biological activity.

It is assumed that extensive management, with higher plant species richness, lower disturbance and reduced tillage, leads to more resilient systems that have the potential to mitigate climate change effects and assure sustainable agricultural productivity (Isbell et al., 2017). Despite the beneficial net effect of extensive management described above, our study revealed that, contrary to our expectations, the extensively-managed systems experienced the greatest losses in soil biological activity. Extensively managed systems had reduced functions under future climate conditions, particularly for soil invertebrate feeding activity. Activity levels in extensive grasslands thus seemed more vulnerable to climate change than in intensive grasslands. These results contradict a large body of literature reporting buffering effects of high-biodiversity systems under environmental stress (Isbell et al., 2015). However, it should also be noted that extensive grasslands showed high soil biological activity under ambient climate conditions, which means that they may have responded more to the climate treatment, because they have 'more to lose' (Pfisterer and Schmid, 2002; Wright et al., 2015).

There is broad evidence that systems with higher plant diversity are better capable of resisting environmental disturbances, e.g., based on their greater range of (asynchronous) responses (Craven et al., 2018; Hector et al., 2010; Mazancourt et al., 2013) and the higher probability of containing tolerant species able to access limited resources (Mueller et al., 2013), thus facilitating the reliable provisioning of ecosystem functions under environmental change (Hautier et al., 2015; Keith et al., 2008; Milcu et al., 2010). However, we could not confirm such plant diversity- and/or management-mediated resilience effects on soil biological activity in our two-year study, in which climate change as well as land use were simulated in realistic scenarios. Despite the evidence of alterations in soil biological activities in our two-year study, it is still difficult to predict how much time those impaired functions may need to recover to pre-treatment levels in the different land-use types. For drawing such conclusions, long-term monitoring of soil biological activities after the cessation of treatments would be required.

What makes our findings particularly interesting is that they are derived from an experimental framework with rather conservative assumptions. First, the climate treatment applied in this GCEF experiment was moderate, at less than +1 °C, when compared to most climate warming experiments and the stated aim of the Paris Agreement to limit temperature rise to below 2 °C in order to avoid the most detrimental climate change effects (Paris Agreement, 2015). Second, the experiment was situated on chernozem soil, which is highly fertile and has a higher water-holding capacity than many other soil types that might be more susceptible to environmental stress (Altermann et al., 2005). Accordingly, future studies should investigate the context-dependency of climate change effects on soil communities and functions. For instance, globally coordinated networks of experiments, such as Drought-Net (Knapp et al., 2017), provide the ideal set-up to explore climate change effects across different climates and soils.

Our results would suggest that the high levels of soil biological activity in extensively-used grasslands are driven by organisms that are particularly vulnerable to environmental stress and whose function provisioning cannot be solely preserved via increases in or the maintenance of plant diversity. In intensive meadows, by contrast, activity levels are presumably driven by a soil community already adapted to disturbances and therefore only little affected by climate change. However, even in non-changed conditions, these intensive meadows reflect the low biological activity level of an already functionally deteriorated ecosystem. As Tsiafouli et al. (2015) demonstrated, intensive management results in a dramatic reduction in soil functioning, including nutrient cycling, decomposition, and natural pest control. This would make the case for adopting practices that promote highly diverse systems, even if they appear to be less resilient to environmental stress (Pfisterer and Schmid, 2002, Wright et al., 2015).

We would argue that new management approaches, besides simply increasing plant diversity, are needed to complement the beneficial effects of extensive management. This could include the selection of specific plant species compositions that are more resistant to drought periods and other climate extremes (Madani et al., 2018). In this vein, related ideas like engineered plant communities (Storkey et al., 2015) or targeted diversified agroecosystems (Isbell et al., 2017) are promising future applied research directions to balance productivity and stability of ecosystem functioning above- and below-ground. Both are based on the notion that a stable provisioning of agroecosystem services (e.g. food production, soil fertility, and pest control) can best if not only be achieved if we foster functionally diverse communities with contrasting traits, e.g., regarding phenology or growth type (Storkey et al., 2015). By implementing diversification strategies, such as increased genetic diversity or crop rotations in agroecosystems, we might be able to overcome declining yields that are predicted for monocultures within the next decades (Isbell et al., 2017). Moreover, such combined approaches might be key to secure ecosystem functioning and food provisioning in the future by supporting systems that hold a vital level of below-ground functionality.

5. Conclusions and outlook

We conclude that climate change consistently reduced soil biological activity throughout the year, without any detectable compensation among seasons. Furthermore, future climate conditions may lead to significant phenological shifts in soil organisms that may cause as yet unexplored community changes and potential mismatches of above- and below-ground interactions (Eisenhauer et al., 2018). Our findings call for future research on the potential context-dependencies of climate-change effects on soil organisms and functions, such as by employing globally distributed experiments covering different environmental conditions (Knapp et al., 2017). Moreover, in the present study soil animal activity showed more pronounced responses to the climate change treatment than our soil microbial activity measurement. Although the applied methods are not straightforward to compare, varying vulnerabilities of different groups of soil organisms should be explored in future studies (George et al., 2019; Siebert et al., 2019).

Our work would corroborate the expectation that extensive management can support higher levels of soil biological activity and related functions than intensive management practices. However, current extensive management practice, as tested here, may not be sufficient to alleviate predicted climate change effects and therefore needs to be complemented by other approaches. We encourage future research to investigate new avenues, such as the use of targeted plant species compositions, to maintain high levels of soil biological activity in agricultural landscapes in a changing world.

Acknowledgements

We thank the staff of the Bad Lauchstädt Experimental Research Station (especially Ines Merbach and Konrad Kirsch) for their work in maintaining the plots and infrastructures of the Global Change Experimental Facility (GCEF), and Harald Auge, François Buscot, and Stefan Klotz for their role in setting up the GCEF. We also thank Sigrid Berger for providing the data on bare soil cover, Hannah Pfeifer for data on root biomass, and Alla Kavtea, Claudia Breitkreuz, Tom Künne, and Ulrich Pruschitzki for their support with lab and field work. MPT acknowledges funding from the German Research Foundation (DFG, TH 2307/1–1). Financial support came from the German Centre for Integrative Biodiversity Research Halle-Jena-Leipzig, funded by the German Research Foundation (FZT 118).

Author contributions

M.S. and T.R. are part of the GCEF steering committee that developed the experimental platform. N.E. conceived the study on soil microbial and invertebrate activity. J.S. and T.R. collected the data. J.S. and M.P.T. analysed the data. J.S. wrote the manuscript with contributions from all authors.

Competing interests

The authors declare no competing interests.

References

A'Bear, A.D., Boddy, L., Hefin Jones, T., 2012. Impacts of elevated temperature on the growth and functioning of decomposer fungi are influenced by grazing collembola. Glob. Chang. Biol. 18, 1823–1832.

Allison, S.D., Treseder, K.K., 2008. Warming and drying suppress microbial activity and carbon cycling in boreal forest soils. Glob. Chang. Biol. 14, 2898–2909.

Altermann, M., Rinklebe, J., Merbach, I., Körschens, M., Langer, U., Hofmann, B., 2005. Chernozem—soil of the year 2005. J. Plant Nutr. Soil Sci. 168, 725–740.

Bakonyi, G., Nagy, P., Kovacs-Lang, E., Kovacs, E., Barabás, S., Répási, V., Seres, A., 2007. Soil nematode community structure as affected by temperature and moisture in a temperate semiarid shrubland. Appl. Soil Ecol. 37, 31–40.

Balvanera, P., Pfisterer, A.B., Buchmann, N., He, J.S., Nakashizuka, T., Raffaelli, D., Schmid, B., 2006. Quantifying the evidence for biodiversity effects on ecosystem functioning and services. Ecol. Lett. 9, 1146–1156.

Bardgett, R.D., Van Der Putten, W.H., 2014. Belowground biodiversity and ecosystem functioning. Nature 515, 505–511.

Bardgett, R.D., Wardle, D.A., 2010. Aboveground-Belowground Linkages: Biotic Interactions, Ecosystem Processes, and Global Change. Oxford University Press Oxford.

Beier, C., Emmett, B., Gundersen, P., Tietema, A., Penuelas, J., Estiarte, M., Gordon, C., Gorissen, A., Llorens, L., Roda, F., 2004. Novel approaches to study climate change effects on terrestrial ecosystems in the field: drought and passive nighttime warming. Ecosystems 7, 583–597.

Bengtsson, J., Ahnström, J., Weibull, A.C., 2005. The effects of organic agriculture on biodiversity and abundance: a meta-analysis. J. Appl. Ecol. 42, 261–269.

Birkhofer, K., Bezemer, T.M., Bloem, J., Bonkowski, M., Christensen, S., Dubois, D., Ekelund, F., Fließbach, A., Gunst, L., Hedlund, K., 2008. Long-term organic farming fosters below and aboveground biota: implications for soil quality, biological control and productivity. Soil Biol. Biochem. 40, 2297–2308.

Birkhofer, K., Diekötter, T., Boch, S., Fischer, M., Müller, J., Socher, S., Wolters, V., 2011. Soil fauna feeding activity in temperate grassland soils increases with legume and grass species richness. Soil Biol. Biochem. 43, 2200–2207.

Birkhofer, K., Bezemer, T.M., Hedlund, K., Setälä, H., 2012. Community composition of soil organisms under different wheat farming systems. In: Cheeke, T.E., Coleman, D.C., Wall, D.H. (Eds.), Microbial Ecology in Sustainable Agroecosystems. Crc Press Taylor & Francis Group, Boca Raton.

Blankinship, J.C., Niklaus, P.A., Hungate, B.A., 2011. A meta-analysis of responses of soil biota to global change. Oecologia 165, 553–565.

Briones, M.J.I., Ineson, P., Heinemeyer, A., 2007. Predicting potential impacts of climate change on the geographical distribution of enchytraeids: a meta-analysis approach. Glob. Chang. Biol. 13, 2252–2269.

Briones, M.J.I., Ostle, N.J., McNamara, N.P., Poskitt, J., 2009. Functional shifts of grassland soil communities in response to soil warming. Soil Biol. Biochem. 41, 315–322.

Brown, J.H., Gillooly, J.F., Allen, A.P., Savage, V.M., West, G.B., 2004. Toward a metabolic theory of ecology. Ecology 85, 1771–1789.

Butenschoen, O., Scheu, S., Eisenhauer, N., 2011. Interactive effects of warming, soil humidity and plant diversity on litter decomposition and microbial activity. Soil Biol. Biochem. 43, 1902–1907.

Cohen, J.M., Lajeunesse, M.J., Rohr, J.R., 2018. A global synthesis of animal phenological responses to climate change. Nat Climate Change 8, 224–228.

Coleman, D.C., Crossley Jr., D.A., Hendrix, P.F., 2004. Fundamentals of Soil Ecology. Academic press.

Conant, R.T., Ryan, M.G., Ågren, G.I., Birge, H.E., Davidson, E.A., Eliasson, P.E., Evans, S.E., Frey, S.D., Giardina, C.P., Hopkins, F.M., 2011. Temperature and soil organic matter decomposition rates–synthesis of current knowledge and a way forward. Glob. Chang. Biol. 17, 3392–3404.

Craven, D., Eisenhauer, N., Pearse, W.D., Hautier, Y., Roscher, C., Boenisch, A.H., Kattge, J., Kreyling, J., Lanta, V., Enrica De Luca, H.W., 2018. Multiple facets of biodiversity drive the diversity-stability relationship. Nat. Ecol. Evol. 2, 1579–1587.

Davidson, E.A., Janssens, I.A., 2006. Temperature sensitivity of soil carbon decomposition and feedbacks to climate change. Nature 440, 165–173.

De Vries, F.T., Liiri, M.E., Bjørnlund, L., Bowker, M.A., Christensen, S., Setälä, H.M., Bardgett, R.D., 2012. Land use alters the resistance and resilience of soil food webs to drought. Nat. Clim. Chang. 2, 276–280.

Decaëns, T., Jiménez, J., 2002. Earthworm communities under an agricultural intensification gradient in Colombia. Plant Soil 240, 133–143.

Delbart, N., Picard, G., Le Toan, T., Kergoat, L., Quegan, S., Woodward, I., Dye, D., Fedotova, V., 2008. Spring phenology in boreal Eurasia over a nearly century time scale. Glob. Chang. Biol. 14, 603–614.

Döring, T.F., Kromp, B., 2003. Which carabid species benefit from organic agriculture?—a review of comparative studies in winter cereals from Germany and Switzerland. Agric. Ecosyst. Environ. 98, 153–161.

Doscher, R., Willén, U., Jones, C., Rutgersson, A., Meier, H.M., Hansson, U., Graham, L.P., 2002. The development of the regional coupled ocean-atmosphere model Rcao. Boreal Environ. Res. 7, 183–192.

Dukes, J.S., Chiariello, N.R., Cleland, E.E., Moore, L.A., Shaw, M.R., Thayer, S., Tobeck, T., Mooney, H.A., Field, C.B., 2005. Responses of grassland production to single and multiple global environmental changes. PLoS Biol. 3 e319.

Durant, J.M., Hjermann, D.Ø., Ottersen, G., Stenseth, N.C., 2007. Climate and the match or mismatch between predator requirements and resource availability. Climate Res. 33, 271–283.

Eisenhauer, N., Dobies, T., Cesarz, S., Hobbie, S.E., Meyer, R.J., Worm, K., Reich, P.B., 2013. Plant diversity effects on soil food webs are stronger than those of elevated CO2 and N deposition in a long-term grassland experiment. Proc. Natl. Acad. Sci. 110 (17), 6889–6894.

Eisenhauer, N., Herrmann, S., Hines, J., Buscot, F., Siebert, J., Thakur, M.P., 2018. The dark side of animal phenology. Trends Ecol. Evol. 33 (12), 898–901.

Fierer, N., Craine, J.M., Mclauchlan, K., Schimel, J.P., 2005. Litter quality and the temperature sensitivity of decomposition. Ecology 86, 320–326.

Foley, J.A., Defries, R., Asner, G.P., Barford, C., Bonan, G., Carpenter, S.R., Chapin, F.S., Coe, M.T., Daily, G.C., Gibbs, H.K., 2005. Global consequences of land use. Science 309, 570–574.

França, S.C., Gomes-Da-Costa, S.M., Silveira, A.P., 2007. Microbial activity and arbuscular mycorrhizal fungal diversity in conventional and organic citrus orchards. Biol. Agric. Hortic. 25, 91–102.

Gardi, C., Montanarella, L., Arrouays, D., Bispo, A., Lemanceau, P., Jolivet, C., Mulder, C., Ranjard, L., Römbke, J., Rutgers, M., 2009. Soil biodiversity monitoring in Europe: ongoing activities and challenges. Eur. J. Soil Sci. 60, 807–819.

George, P.B., Lallias, D., Creer, S., Seaton, F.M., Kenny, J.G., Eccles, R.M., Griffiths, R.I., Lebron, I., Emmett, B.A., Robinson, D.A., 2019. Divergent national-scale trends of microbial and animal biodiversity revealed across diverse temperate soil ecosystems. Nat. Commun. 10, 1107.

Giller, K., Beare, M., Lavelle, P., Izac, A.-M., Swift, M., 1997. Agricultural intensification, soil biodiversity and agroecosystem function. Appl. Soil Ecol. 6, 3–16.

Gillooly, J.F., Brown, J.H., West, G.B., Savage, V.M., Charnov, E.L., 2001. Effects of size and temperature on metabolic rate. Science 293, 2248–2251.

Goldenberg, S.U., Nagelkerken, I., Marangon, E., Bonnet, A., Ferreira, C.M., Connell, S.D., 2018. Ecological complexity buffers the impacts of future climate on marine consumers. Nat. Clim. Change 8, 229–233.

Hamel, C., Schellenberg, M.P., Hanson, K., Wang, H., 2007. Evaluation of the "bait-lamina test" to assess soil microfauna feeding activity in mixed grassland. Appl. Soil Ecol. 36, 199–204.

Harding, D., Stuttard, R., 1974. Microarthropods. In: Dickinson, C.H., Pugh, G.J.F. (Eds.), Biology of Plant Litter Decomposition. Academic Press, pp. 489–532.

Hautier, Y., Tilman, D., Isbell, F., Seabloom, E.W., Borer, E.T., Reich, P.B., 2015. Anthropogenic environmental changes affect ecosystem stability via biodiversity. Science 348, 336–340.

Hector, A., Hautier, Y., Saner, P., Wacker, L., Bagchi, R., Joshi, J., Scherer-Lorenzen, M., Spehn, E.M., Bazeley-White, E., Weilenmann, M., 2010. General stabilizing effects of plant diversity on grassland productivity through population asynchrony and overyielding. Ecology 91, 2213–2220.

Ingram, J., Gregory, P., Izac, A.-M., 2008. The role of agronomic research in climate change and food security policy. Agr Ecosyst Environ 126, 4–12.

Isbell, F., Craven, D., Connolly, J., Loreau, M., Schmid, B., Beierkuhnlein, C., Bezemer, T.M., Bonin, C., Bruelheide, H., De Luca, E., 2015. Biodiversity increases the resistance of ecosystem productivity to climate extremes. Nature 526, 574–577.

Isbell, F., Adler, P.R., Eisenhauer, N., Fornara, D., Kimmel, K., Kremen, C., Letourneau, D.K., Liebman, M., Polley, H.W., Quijas, S., Scherer-Lorenzen, M., 2017. Benefits of increasing plant diversity in sustainable agroecosystems. J. Ecol. 105, 871–879.

Jacob, D., Podzun, R., 1997. Sensitivity studies with the regional climate model REMO. Meteorol. Atmos. Phys. 63, 119–129.

Keith, A.M., Van Der Wal, R., Brooker, R.W., Osler, G.H., Chapman, S.J., Burslem, D.F., Elston, D.A., 2008. Increasing litter species richness reduces variability in a terrestrial decomposer system. Ecology 89, 2657–2664.

Kerr, R.A., 2007. Global warming is changing the world. Science 316, 188–190.

Knapp, A.K., Avolio, M.L., Beier, C., Carroll, C.J., Collins, S.L., Dukes, J.S., Fraser, L.H., Griffin-Nolan, R.J., Hoover, D.L., Jentsch, A., 2017. Pushing precipitation to the extremes in distributed experiments: recommendations for simulating wet and dry years. Glob. Chang. Biol. 23, 1774–1782.

Kratz, W., 1998. The bait-lamina test. Environ. Sci. Pollut. Res. 5, 94–96.

Lange, M., Eisenhauer, N., Sierra, C.A., Bessler, H., Engels, C., Griffiths, R.I., Mellado-Vázquez, P.G., Malik, A.A., Roy, J., Scheu, S., 2015. Plant diversity increases soil microbial activity and soil carbon storage. Nat. Commun. 6, 6707.

Loreau, M., Mouquet, N., Gonzalez, A., 2003. Biodiversity as spatial insurance in heterogeneous landscapes. Proc. Natl. Acad. Sci. 100, 12765–12770.

Macfadyen, A., 1961. Improved funnel-type extractors for soil arthropods. J. Anim. Ecol. 30, 171–184.

Madani, N., Kimball, J.S., Ballantyne, A.P., Affleck, D.L., Bodegom, P.M., Reich, P.B., Kattge, J., Sala, A., Nazeri, M., Jones, M.O., 2018. Future global productivity will be affected by plant trait response to climate. Sci. Rep. 8, 2870.

Manzoni, S., Schimel, J.P., Porporato, A., 2012. Responses of soil microbial communities to water stress: results from a meta-analysis. Ecology 93, 930–938.

Mazancourt, C., Isbell, F., Larocque, A., Berendse, F., Luca, E., Grace, J.B., Haegeman, B., Wayne Polley, H., Roscher, C., Schmid, B., 2013. Predicting ecosystem stability from community composition and biodiversity. Ecol. Lett. 16, 617–625.

Meinke, I., Gerstner, E., von Storch, H., Marx, A., Schipper, H., Kottmeier, C., Treffeisen, R., Lemke, P., 2010. Regionaler klimaatlas deutschland der Helmholtz-Gemeinschaft informiert im Internet über möglichen künftigen Klimawandel. Mitt. DMG 2, 5–7.

Melillo, J., Steudler, P., Aber, J., Newkirk, K., Lux, H., Bowles, F., Catricala, C., Magill, A., Ahrens, T., Morrisseau, S., 2002. Soil warming and carbon-cycle feedbacks to the climate system. Science 298, 2173–2176.

Menzel, A., Fabian, P., 1999. Growing season extended in Europe. Nature 397, 659.

Milcu, A., Thebault, E., Scheu, S., Eisenhauer, N., 2010. Plant diversity enhances the reliability of belowground processes. Soil Biol. Biochem. 42, 2102–2110.

Moreno, B., Garcia-Rodriguez, S., Cañizares, R., Castro, J., Benítez, E., 2009. Rainfed olive farming in South-Eastern Spain: long-term effect of soil management on biological indicators of soil quality. Agric. Ecosyst. Environ.rgg 131, 333–339.

Mueller, K.E., Tilman, D., Fornara, D.A., Hobbie, S.E., 2013. Root depth distribution and the diversity–productivity relationship in a long-term grassland experiment. Ecology 94, 787–793.

Newbold, T., Hudson, L.N., Hill, S.L., Contu, S., Lysenko, I., Senior, R.A., Börger, L., Bennett, D.J., Choimes, A., Collen, B., 2015. Global effects of land use on local terrestrial biodiversity. Nature 520, 45–50.

Nord, E.A., Lynch, J.P., 2009. Plant phenology: a critical controller of soil resource acquisition. J. Exp. Bot. 60, 1927–1937.

Paris Agreement, 2015. United Nations Framework Convention on Climate Change. Paris, France.

Peñuelas, J., Filella, I., 2009. Phenology feedbacks on climate change. Science 324, 887–888.

Pfisterer, A.B., Schmid, B., 2002. Diversity-dependent production can decrease the stability of ecosystem functioning. Nature 416, 84.

Pinheiro J., Bates D., DebRoy S., Sarkar D., R Core Team (2017). nlme: Linear and Nonlinear Mixed Effects Models. R package version 3.1-128

R Core Team, R. C. T, 2017. R: A Language and Environment for Statistical Computing. Vienna, Austria; 2014.

Rockel, B., Will, A., Hense, A., 2008. The regional climate model COSMO-CLM (CCLM). Meteorol. Z. 17, 347–348.

Rożen, A., Sobczyk, Ł., Liszka, K., Weiner, J., 2010. Soil faunal activity as measured by the bait-lamina test in monocultures of 14 tree species in the Siemianice common-garden experiment, Poland. Appl. Soil Ecol. 45, 160–167.

Rustad, L., Campbell, J., Marion, G., Norby, R., Mitchell, M., Hartley, A., Cornelissen, J., Gurevitch, J., 2001. A meta-analysis of the response of soil respiration, net nitrogen mineralization, and aboveground plant growth to experimental ecosystem warming. Oecologia 126, 543–562.

Ryan, J.A., Ulrich, J.M., Thielen, W., Teetor, P., Ulrich, M.J.M., 2017. Package 'quantmod'.

Sala, O.E., Chapin, F.S., Armesto, J.J., Berlow, E., Bloomfield, J., Dirzo, R., Huber-Sanwald, E., Huenneke, L.F., Jackson, R.B., Kinzig, A., 2000. Global biodiversity scenarios for the year 2100. Science 287, 1770–1774.

Sánchez-Moreno, S., Castro, J., Alonso-Prados, E., Alonso-Prados, J.L., García-Baudín, J.M., Talavera, M., Durán-Zuazo, V.H., 2015. Tillage and herbicide decrease soil biodiversity in olive orchards. Agron. Sustain. Dev. 35, 691–700.

Schädler, M., Buscot, F., Klotz, S., Reitz, T., Durka, W., Bumberger, J., Merbach, I., Michalski, S.G., Kirsch, K., Remmler, P., Schulz, E., Auge, H., 2019. Investigating the Consequences of Climate Change under Different Land-Use Regimes—A Novel Experimental Infrastructure. Ecosphere 10 (3), e02635.

Scheu, S., 1992. Automated measurement of the respiratory response of soil micro-compartments: active microbial biomass in earthworm faeces. Soil Biol. Biochem. 24, 1113–1118.

Schulz, E., 2002. Influence of extreme management on decomposable soil organic matter pool. Arch. Agron. Soil Sci. 48, 101–105.

Siebert, J., Suennemann, M., Auge, H., Berger, S., Cesarz, S., Ciobanu, M., Guerrero-Ramirez, N.R., Eisenhauer, N., 2018. The effects of drought and nutrient addition on soil organisms vary across taxonomic groups, but are constant across seasons. Sci. Rep. 9 (1), 639.

Simpson, J.E., Slade, E., Riutta, T., Taylor, M.E., 2012. Factors affecting soil fauna feeding activity in a fragmented lowland temperate deciduous woodland. PLoS One 7, e29616.

Storkey, J., Döring, T., Baddeley, J., Collins, R., Roderick, S., Jones, H., Watson, C., 2015. Engineering a plant community to deliver multiple ecosystem services. Ecol. Appl. 25, 1034–1043.

Swift, M.J., Heal, O.W., Anderson, J.M., 1979. Decomposition in Terrestrial Ecosystems. Univ of California Press.

Thackeray, S.J., Henrys, P.A., Hemming, D., Bell, J.R., Botham, M.S., Burthe, S., Helaouet, P., Johns, D.G., Jones, I.D., Leech, D.I., 2016. Phenological sensitivity to climate across taxa and trophic levels. Nature 535, 241.

Thakur, M.P., Reich, P.B., Hobbie, S.E., Stefanski, A., Rich, R., Rice, K.E., Eddy, W.C., Eisenhauer, N., 2018. Reduced feeding activity of soil detritivores under warmer and drier conditions. Nat. Clim. Change 8, 75–78.

Thomson, L.J., Macfadyen, S., Hoffmann, A.A., 2010. Predicting the effects of climate change on natural enemies of agricultural pests. Biol. Control 52, 296–306.

Tilman, D., Cassman, K.G., Matson, P.A., Naylor, R., Polasky, S., 2002. Agricultural sustainability and intensive production practices. Nature 418, 671–677.

Treseder, K.K., 2008. Nitrogen additions and microbial biomass: a meta-analysis of ecosystem studies. Ecol. Lett. 11, 1111–1120.

Tsiafouli, M.A., Thébault, E., Sgardelis, S.P., Ruiter, P.C., Putten, W.H., Birkhofer, K., Hemerik, L., Vries, F.T., Bardgett, R.D., Brady, M.V., 2015. Intensive agriculture reduces soil biodiversity across Europe. Glob. Chang. Biol. 21, 973–985.

van Rij, J., Wieling, M., Baayen, R., van Rijn, H., 2017. itsadug: Interpreting Time Series and Autocorrelated Data Using GAMMs. R package version 2.3.

Verhoef, H., Brussaard, L., 1990. Decomposition and nitrogen mineralization in natural and agroecosystems: the contribution of soil animals. Biogeochemistry 11, 175–211.

Vitousek, P.M., 1994. Beyond global warming: ecology and global change. Ecology 75, 1861–1876.

Walter, J., Hein, R., Beierkuhnlein, C., Hammerl, V., Jentsch, A., Schädler, M., Schuerings, J., Kreyling, J., 2013. Combined effects of multifactor climate change and land-use on decomposition in temperate grassland. Soil Biol. Biochem. 60, 10–18.

Wardle, D., Bonner, K., Barker, G., 2002. Linkages between plant litter decomposition, litter quality, and vegetation responses to herbivores. Funct. Ecol. 16, 585–595.

Wood, S., Scheipl, F., 2017. gamm4: Generalized Additive Mixed Models using 'mgcv' and 'lme4'. R package version 0.2-5.

Wright, A.J., Ebeling, A., De Kroon, H., Roscher, C., Weigelt, A., Buchmann, N., Buchmann, T., Fischer, C., Hacker, N., Hildebrandt, A., 2015. Flooding disturbances increase resource availability and productivity but reduce stability in diverse plant communities. Nat. Commun. 6, 6092.

Yachi, S., Loreau, M., 1999. Biodiversity and ecosystem productivity in a fluctuating environment: the insurance hypothesis. Proc. Natl. Acad. Sci. 96, 1463–1468.

> CHAPTER THREE

Assessing the resilience of biodiversity-driven functions in agroecosystems under environmental change

Emily A. Martin[a],*, Benjamin Feit[b], Fabrice Requier[a], Hanna Friberg[c], Mattias Jonsson[b]

[a]Department of Animal Ecology and Tropical Biology, Biocenter, University of Würzburg, Würzburg, Germany
[b]Department of Ecology, Swedish University of Agricultural Sciences, Uppsala, Sweden
[c]Department of Forest Mycology and Plant Pathology, Swedish University of Agricultural Sciences, Uppsala, Sweden
*Corresponding author: e-mail address: emily.martin@uni-wuerzburg.de

Contents

Advances in Ecological Research, Volume 60
ISSN 0065-2504
https://doi.org/10.1016/bs.aecr.2019.02.003

Abstract

Predicting the resilience of biodiversity-driven functions in agroecosystems to drivers of environmental change (EC) is of critical importance to ensure long-term and environmentally safe agricultural production. However, operationalizing resilience of such functions is challenging, because conceptual approaches differ, direct measures are difficult, and the validity and interpretation of existing indicators are unclear. Here, we (1) summarize dimensions of resilience that apply in agroecosystems, and the disturbances they are subject to under EC. We then (2) review indicators of the resilience of biodiversity-driven functions in agroecosystems, and their support in theoretical and empirical studies. (3) Using these indicators, we examine what can be learned for the resilience of these functions to drivers of EC, focussing on the ecosystem services of biological pest control, biological disease control in soil and pollination. We conclude (4) that research into the resilience of these services is still in its infancy, but novel tools and approaches can catalyse further steps to assess and improve the resilience of biodiversity-driven agroecosystem functions under EC.

1. Introduction

Characterizing and supporting the ability of Earth's ecosystems and their functions to persist, recover and adapt in the face of environmental change (EC) is a major research agenda of the 21st century (Steffen et al., 2018). In agroecosystems, human-environment interactions drive the provision of functions and services—such as crop yields—that are essential for the maintenance of contemporary human societies. However, just as other ecosystems worldwide, agroecosystems are subject to intensifying drivers of environmental change, which are likely to affect their ability to maintain functions over the long term (Tylianakis et al., 2008). Anticipating and preventing the loss of function in agroecosystems is made pressing by immediate, cascading, and potentially catastrophic impacts on global human food security, health, energy, and socio-economic stability as well as on the broader environment (Cabell and Oelofse, 2012; Foley et al., 2011; Wheeler and von Braun, 2013).

Key functions provided by agroecosystems include pollination and biological control of pests and diseases. Because they directly influence several aspects of crop productivity, these functions represent ecosystem services with particularly critical impacts on human well-being, most notably human food security (Klein et al., 2007; Millennium Ecosystem Assessment, 2005; Oerke, 2006). In contrast to functions mainly driven by interactions between crops and their abiotic environment (e.g. water regulation, soil retention), pollination, pest and disease control are strongly biodiversity-driven, i.e., the presence and structure of service-providing communities such as pollinators and natural enemies determine service provision, contribution to crop yield and nutritional quality of crops (Bommarco et al., 2012; Pywell et al., 2015). In addition to their key contribution to yields, pest and disease control by natural enemies or antagonistic microorganisms represents an alternative management strategy that could limit environmental externalities and negative feedback loops associated with the use and over-use of synthetic pesticides in agriculture (Bommarco et al., 2013; Lechenet et al., 2017). However, while our understanding of the patterns of biodiversity underlying these functions is steadily increasing, the scientific field of predicting biodiversity-driven services in agroecosystems is in many respects still in its infancy (Karp et al., 2018; Tscharntke et al., 2016).

Resilience represents the ability of a system to maintain or recover its functioning, structure and overall identity in the face of changes in environmental conditions (Fig. 1) (Folke et al., 2010; Walker et al., 2004). If resilience of a system is low, disturbances may cause it to pass a threshold or tipping point, after which nonlinear regime shifts may occur (Scheffer et al., 2015). Though often elusive, approaches to quantify and predict the resilience and proximity to tipping points of ecological and social-ecological systems under EC have been fast developing in systems from aquatic, to rangelands, to global plant-pollinator networks (Dakos et al., 2015; Jiang et al., 2018; Sasaki et al., 2015; Sterling et al., 2017). In agroecosystems, such approaches (Angeler and Allen, 2016; Döring et al., 2013; Oliver et al., 2015a; Peterson et al., 2018; Standish et al., 2014) are however only sparsely reflected in the available literature (Vandermeer, 2011). For example, in a review of studies on ecological thresholds of change (Sasaki et al., 2015), agroecosystems were represented in less than 2% (2/147) of all studies published until 2013. More broadly, while resilience research in agroecosystems has focussed to some extent on the maintenance of yields at the field scale (e.g. Döring et al., 2013; Peterson et al., 2018), few studies have explicitly examined the resilience of biodiversity-driven functions and services to contemporary drivers of EC (Donohue et al., 2016; Oliver et al., 2015a). To date, the great majority of studies have used the effects of disturbance on farmland biodiversity as proxies to evaluate the resilience of biodiversity-driven functions (e.g. Karp et al., 2011; Oliver et al., 2015b). However, despite strong theoretical underpinnings and practical benefits of such proxies, the link between actual resilience of agroecosystem functions and the variety of resilience indicators based on biodiversity or other aspects have rarely been demonstrated (Angeler and Allen, 2016; Egli et al., 2018).

In the present review, we synthesize the approaches taken to evaluate the resilience of functioning in agroecosystems and their implications for the maintenance and vulnerability of agroecosystem functioning under EC. We focus on the biodiversity-driven functions and ecosystem services of pollination, biological pest control and biological disease control in soils. After (1) defining the concept of resilience as applied to agroecosystems and the nature and scales of disturbances that affect them, we (2) identify how resilience has been measured in these systems and review which available indicators are demonstrated to link to the resilience of biodiversity-driven functions in agroecosystems (from here on termed agroecosystem functions). We then (3) provide a narrative review of what can be learned from available measures and indicators before (4) highlighting current challenges and novel approaches with the potential to push resilience assessment of agroecosystem functions beyond its present state-of-the-art. Overall, we aim for this review to catalyse the development and implementation of rigorous strategies to understand, manage and predict the resilience of agroecosystem functions under EC, accounting for its multiple dimensions and spatiotemporal complexity.

2. The concept of resilience as applied to biodiversity-driven functions in agroecosystems

2.1 Definitions

Resilience has been defined in a multitude of ways across disciplines and systems (Carpenter et al., 2001; Peterson et al., 2018) leading to considerable breadth for its practical implementation (Donohue et al., 2016; Mori, 2016). To operationalize resilience in the context of this review, we consider three related meanings that have been applied or are relevant in agroecological science (Fig. 1) (Angeler and Allen, 2016; Donohue et al., 2016; Egli et al., 2018). (i) *Persistence* is the ability of a system to maintain its function under stress. (ii) *Engineering resilience* is the ability of a system to return to its previous state (bounce back) after a disturbance, as reflected, e.g., by its speed (or rate) of recovery to the previous state. (iii) *Ecological resilience* is the extent of disturbances a system can absorb before reorganizing into a different state with different functioning, structure, identity and feedbacks (see also Folke et al., 2010; Walker et al., 2004 for a detailed definition of ecological resilience).

These definitions differ in their focus on (i) the maintenance of an acceptable degree of functioning under stress, where time may or may not be explicitly incorporated; (ii) the recovery of functioning after stress, with emphasis on a temporal dimension; and (iii) the existence of threshold disturbance levels associated with the intensity of stress. All three definitions have implications for the measurement and applicability of resilience in agroecosystem management (Sasaki et al., 2015), for the interpretation of underlying mechanisms, and lead to widely differing approaches in empirical and theoretical research. We thus employ these definitions throughout and classify existing approaches into each dimension of resilience.

Resilience (persistence, engineering or ecological resilience) can be brought about by a system or a function's properties of recovery (the ability to bounce back, e.g., by reorganizing after disturbance) or resistance (the ability to withstand disturbance without change; Egli et al., 2018; Hodgson et al., 2015; Oliver et al., 2015a). In some cases, rapid recovery may be interpreted as resistance if observed at wider time steps than the speed of recovery (Oliver et al., 2015a). The extent to which a function is able to *persist* under disturbance depends on both its properties of recovery and resistance (Fig. 1A). In contrast, a function's *engineering resilience* depends on its ability to recover from disturbance (Fig. 1B). Both definitions are linked to the concept of ecological stability under its definition by Holling (1973) and in its extensive use in the literature on, e.g., diversity-stability relationships (Mori, 2016). However, ecological stability also includes the notion of the constancy of a system in space or time (i.e. its lack of variability; Donohue et al., 2016; Grimm and Wissel, 1997). To refer to systems that vary in space or time, we here use the term 'variability' (as opposed to 'invariability' for a system that does not change; Egli et al., 2018). Importantly, in many cases and especially changing agricultural mosaics with heterogeneous patterns of crop type and growth (Vasseur et al., 2013), variability of functions in space or time may be precisely what allows them to recover from and/or resist disturbances (Mori, 2016), thereby exhibiting one or multiple dimensions of resilience (Angeler and Allen, 2016; Egli et al., 2018).

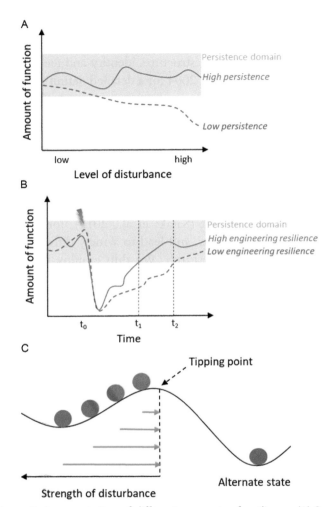

Fig. 1 Schematic representation of different concepts of resilience. (A) Persistence as the ability to continue function provision under high disturbance levels. The system represented by the dashed red line is less resilient than the system shown by the solid blue line. The persistence domain (grey) indicates the amount of function necessary to continue service provision. (B) Engineering resilience as the ability of a system to bounce back after disturbance at t_0. The system represented by the solid blue line is able to recover faster (returning to the persistence domain (grey) at t_1), than the system shown by the dashed red line (returning to the persistence domain at t_2). (C) Ecological resilience as the strength of disturbance a system can absorb before it moves to an alternate state. Illustrated are four different resilience regimes ranging from high (blue) to low resilience (red). The difference in resilience is a consequence of (1) the topography of the domain and (2) the precariousness of the system, i.e., its closeness to the tipping point to an alternate state prior to the disturbance event. The strength of the disturbance required to push the system over the tipping point to an alternate state increases with resilience of the system, indicated by the red arrows. *Adapted with modification from Döring, T.F., Vieweger, A., Pautasso, M., Vaarst, M., Finckh, M.R., Wolfe, M.S., 2013. Resilience as a universal criterion of health. J. Sci. Food Agric. 95, 455–465. https://doi.org/10.1002/jsfa.6539.*

2.2 Spatial, temporal and organizational scales of disturbance: Resilience *to* what?

Disturbance is one of the key features of many agroecosystems, which can be seen as mosaics of repeatedly and differentially disturbed patches through space and time. In agroecosystems, disturbance takes place from the scale of soil aggregates to whole landscapes and biogeographical realms; from instantaneous to decade-long effects; and from individual organisms to whole communities and networks. The multiscale nature of disturbances, and more generally of the variables influencing functions in agroecosystems, means that any given system is affected simultaneously by fast (small-scale) variables, and by slow (large-scale) ones that change much more gradually. Formally, disturbances can be characterized in addition to their spatiotemporal scale in terms of magnitude, frequency, duration, and variability or directionality of change in space and time (Donohue et al., 2016). 'Pulses' or 'acute' disturbances occur more or less instantaneously, but may be distributed over time as discrete environmental fluctuations. 'Press' or 'chronic' disturbances represent sustained, long-term changes (Donohue et al., 2016; Peterson et al., 2018; Sasaki et al., 2015).

Similar to other systems, agroecosystems are affected by globally relevant 'fast' and 'slow' variables (Biggs et al., 2015), such as competition by invasive species, extreme climate events or gradual temperature increases under climate change (e.g. as reviewed by Sasaki et al., 2015). However, a wide range of disturbances relates particularly to agricultural intensification and land use change, with scales that are highly specific to agroecosystems and their management (Peterson et al., 2018). In Fig. 2, we summarize the types of disturbance that can occur in agroecosystems and their spatial and temporal scale of effect. We focus on disturbances whose nature, magnitude or frequency are under the direct influence of physical drivers of terrestrial EC as defined by, e.g., Sala et al. (2000), including particularly disturbances associated with land use change, climate change, and biotic invasions. As such, these disturbances reflect the impact of EC drivers at the level of agroecosystems. Their effects on biodiversity and associated functions occur both within fields and across whole landscapes, and include disturbances associated with direct management of fields and habitats (tilling, harvesting, pesticide application, changes in crops planted and rotations, changes in field sizes), as well as less controlled (less predictable) events such as extreme weather, outbreaks of pests and diseases or other invasions. Under EC, the directionality of these disturbances may be fixed (e.g. global temperature

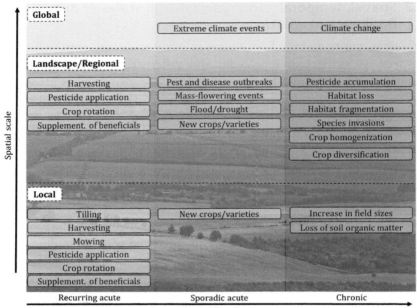

Fig. 2 Types of disturbances characteristic of agroecosystems that affect biodiversity-driven functions across spatial (on the y-axis), temporal and organizational scales. Disturbances listed here refer particularly to drivers of global environmental change including land use change, climate change and biotic invasions. Regularly recurring, anthropogenic 'pulses' (recurring acute disturbances) are key features of most agroecosystems that may intensify or decrease under agricultural land use change. They take place at the same time as other, non-recurring pulses (sporadic acute disturbances such as extreme climate events) and 'presses' (chronic disturbances), which represent gradual changes taking place over long periods of time or large spatial scales. Local, field-level disturbances (crop rotation, introduction of new crops & varieties, supplementation of beneficial species including pollinators and antagonists, pesticide application, harvesting, changes in field size) frequently upscale to impact biodiversity-driven functions at the landscape level.

increases) or may be subject to differences according to regional factors and regulations, as in the case of opposing trends of land use intensification vs. abandonment in different regions and systems (van Vliet et al., 2015).

3. The link between indicators and resilience of agroecosystem functions

To really understand the resilience of agroecosystem functions, we would ideally want to assess both resilience of the function itself and the biodiversity underpinning it in relation to a disturbance. In some situations this is feasible, such as when assessing persistence against certain acute or chronic disturbances occurring at local or landscape scales (Fig. 2). However, in other cases, due to inherent difficulties of measuring the resilience of functions (see Section 5 below; Egli et al., 2018), indicators based on the state of biodiversity have been employed to infer the resilience of functioning. Indeed, developing sets of surrogates or indicators are often seen as a more practicable approach to assessing resilience 'than trying to measure resilience itself' (Cabell and Oelofse, 2012; Darnhofer et al., 2010).

In this section, we summarize major indicators of resilience for biodiversity-driven functions in agroecosystems. For each indicator, we examine how and why it should affect function resilience and to what extent this link has been demonstrated in empirical and theoretical literature. We identify two broad categories of indicators for biodiversity-driven function resilience (Table 1): (1) indicators based on the state of measures of biodiversity, (2) indicators based not on biodiversity, but on statistical or structural properties of agroecosystems and agroecosystem functions.

3.1 Indicators of resilience based on measures of biodiversity

3.1.1 Species richness

Biodiversity-based indicators of function resilience are founded on the premise that components of biodiversity influence how associated functions are affected by disturbances. Among these, the *species richness* of communities has historically been examined as a key indicator of resilience of ecosystem functioning (Hooper et al., 2005, 2012). The idea that species richness can contribute to the resilience of ecosystem functions under disturbance or global change is based on the concept of a diversity of responses to disturbance among species (Elmqvist et al., 2003). Due to species-specific responses to disturbance, individual species within a community are likely to be affected differently by environmental change, depending, for example, on climatic tolerance, drought resistance, or resource requirements (Chapin et al., 1997; Naeem and Wright, 2003). With every additional species in a community, the likelihood increases that some species continue to provide a service when others are lost or reduced in effectiveness because of changes in their environment, through a statistical 'portfolio' effect (see functional redundancy below; Biggs et al., 2015; Peterson et al., 1998). The importance of high species richness for function resilience has been demonstrated in empirical and theoretical settings most frequently for plant productivity (Balvanera et al., 2006; Hooper et al., 2005; Isbell et al., 2015; Yachi and Loreau, 1999). However, recognition has risen that species richness per se does not imply function resilience, because species-rich systems may have

Table 1 Indicators of agroecosystem function resilience and theoretical and empirical evidence for their use.

Indicator	Definition	How does the indicator affect resilience of functioning?	Resilience property affected	Evidence for a link between indicator and function resilience	
				Models	Data
Indicators of function resilience based on measures of biodiversity					
Species richness or diversity (Elmqvist et al., 2003; Peterson et al., 1998)	Number or diversity of species in a community	'Portfolio effect' contributing to the insurance value of biodiversity. But, species-rich systems may still be very vulnerable to disturbance, and ignores the fact that loss of species if non-random	*Resistance or recovery* (cf. mechanisms of response diversity)	High species richness leads to persistence of functions by buffering disturbances (e.g. Hooper et al., 2005; Yachi and Loreau, 1999)	Found for some but not all disturbances (e.g. not drought, warming or high environmental variance) (Balvanera et al., 2006)* (*various functions*); Isbell et al., 2015 (*plant productivity*)
Presence of key species or functional groups (Elmqvist et al., 2003; Folke et al., 2004; Gagic et al., 2015)	Presence of species or functional groups providing the majority of a function	Presence of key groups is necessary to provide function; their loss leads to loss of function if key functional group is removed	*Resistance*; regime shift to 'no function' if key functional group is removed	The consequences of the number of species removed from a system depend on the function those species perform in the system (Dunne et al., 2002)	Maas et al., 2015* (*pest control*); Caves et al., 2013 (*seed dispersal*); Anderson et al., 2011; Garibaldi et al., 2009* (*pollination*); Weller et al., 2002 (*soil disease control*)
Functional diversity (Elmqvist et al., 2003; Folke et al., 2004; Gagic et al., 2015)	Diversity of functions performed by organisms in the community	May lead to presence of key groups (see above) through sampling effects; may enhance function continuity in spatiotemporal crop mosaics; implies high niche occupation, which should buffer against invasion and altered interactions	*Resistance* of functioning due to continued performance in different crops/times or due to competition for niche space		Eisenhauer et al., 2013; van Elsas et al., 2012* (*soil disease control*); Fargione and Tilman, 2005 (*plant productivity*)
Functional redundancy (Biggs et al., 2015; De Bello et al., 2009; Folke et al., 2004; Naeem, 1998; Peterson et al., 1998)	Number of species performing the same function; the fraction of species diversity not expressed by functional diversity	When combined with response diversity within functional groups (see below), functional redundancy leads to increased function resilience through species complementarity	*Resistance or recovery* (cf. mechanisms of response diversity)	Functional redundancy leads to function resilience (Naeem, 1998) and decreases extinction risk of functional groups (Fonseca and Ganade, 2001)	Hallett et al., 2017 (*pollination*); Joner et al., 2011 find no effects on community but not function resilience (*plant productivity*); Sanders et al., 2018* (*pest control*); Griffiths and Philippot, 2013* (*soil functions*)
Response diversity (Biggs et al., 2015; Elmqvist et al., 2003; Mori et al., 2013; Oliver et al., 2015a)	Different species (or individuals within species) contributing to the same function respond differently to disturbance	Species or individuals performing the same function but responding differently to disturbance compensate for each other under a given disturbance through functional compensation and spatiotemporal complementarity	*Resistance or recovery* of function due to performance of species less or not affected by the disturbance, and/or able to adapt and recover from the disturbance	Norberg et al. (2001) show higher resistance of functioning under disturbance when response diversity is high; other work also points in this direction (reviewed by Hooper et al., 2005)	Cariveau et al., 2013* find no link with persistence of pollination under land use change, but Rader et al., 2013* and Stavert et al., 2018* find buffering under climate and land use change, respectively (*pollination*); Farwig et al., 2017; González-Varo et al., 2017; Moreira et al., 2017 (*seed dispersal*)

Mechanism	Definition	Relationship to function resilience	Resistance/Recovery	Examples and evidence
Cross-scale resilience (Allen et al., 2014; Biggs et al., 2015; Elmqvist et al., 2003; Peterson et al., 1998)	Species or functional groups respond to disturbance at different spatial or temporal scales	Risk-spreading across scales similarly to response diversity; due to nestedness of scales, species can be recruited from, e.g., regional pools to perform functions at the landscape or local scales	*Resistance* due to species unaffected by the scale of disturbance. *Recovery* through recruitment of species at an unaffected scale	Jonsson et al. (2015)* find evidence that pest control in crops is less affected by local disturbance when the surrounding landscape is complex (intermediate landscape hypothesis) (*pest control*); Nash et al., 2016 (*coral refs*)
Negative co-variation (Elmqvist et al., 2003)	The abundance of pairs of species with similar effect on function tends to be negatively correlated across disturbance gradients	Arises through competition in the same functional group (density compensation) or differing responses to disturbance. Similarly to response diversity, leads to species with similar function replacing each other under disturbance	*Resistance* of function due to performance of competitors or species with opposing response to disturbance	In theory leads to higher function persistence similarly to response diversity and is associated with the statistical portfolio effect of diversity (Hooper et al., 2005)
Relationship between effect and response traits (Díaz et al., 2013; Lavorel and Garnier, 2002; Oliver et al., 2015a; Suding et al., 2008)	Degree to which effects of species on function (their *effect* traits) are correlated with the responses of species to disturbance (their *response* traits)	Correlation between effect and response traits implies high risk of loss of function if a response group is removed. Conversely, if traits are unrelated, function persistence is decoupled from the response of single groups to disturbance	*Resistance* of functioning through maintenance of effect traits across different levels or types of disturbance	Larsen et al. (2005)* show unexpectedly large consequences for function persistence of non-random species loss in relation to their function (*pollination, dung burial*). Díaz et al. (2013) assess the theoretical risk of loss of function due to trait correlation in five case studies (*decomposition, bushmeat harvest rate, seed dispersal*)
Presence, redundancy and response diversity of biological legacies and/or mobile link organisms (Elmqvist et al., 2003; Folke et al., 2004; Lundberg and Moberg, 2003)	Biological legacies remain in the system after disturbance and provide sources of regrowth (e.g. seed bank, eggs in diapause). Mobile links are able to recolonize patches after a disturbance	(1) Abundance, survival and growth rate of biological legacies and mobile links; (2) their support by, e.g., undisturbed areas in the surrounding landscape; and (3) the accessibility of disturbed patches for mobile link organisms (i.e. their dispersal abilities), determine the speed at which these species can recolonize and provide functions in disturbed patches	*Recovery* of function after disturbance through renewal and reorganization	Seed dispersal resilience is enhanced by mobile link birds and by biological legacies (isolated trees and forest patches), but not by bird response diversity and functional redundancy (García et al., 2013) (*seed dispersal*)
Network structure and interactions (Biggs et al., 2015; Folke et al., 2004; Tylianakis and Morris, 2017)	Strength, number and structure of interactions between species of a community	Function resilience should increase with resource use complementarity and/or nestedness or modularity of a network, due to disturbances only affecting (modifying or removing) a subset of existing interactions; network connectedness however is not necessarily positive for resilience of associated functions	*Resistance and recovery* take place through buffering by remaining network links after disturbance and through formation of new links, respectively	Network complementarity affects functioning under species loss (Poisot et al., 2013); nestedness improves network recovery (Okuyama and Holland, 2008; Thébault and Fontaine, 2010); connectance, nestedness affect proneness to sudden collapse of pollinators (Lever et al., 2014). Peralta et al. (2014) show network complementarity enhances parasitism and decreases its spatial variability, but no test of persistence under disturbance; apparent competition shapes response to change of persistence (Frost et al., 2016) (*pest control*)

Continued

Table 1 Indicators of agroecosystem function resilience and theoretical and empirical evidence for their use.—cont'd

Indicator	Definition	How does the indicator affect resilience of functioning?	Resilience property affected	Evidence for a link between indicator and function resilience	
				Models	Data
Other indicators of function resilience					
Recovery rate (Scheffer et al., 2015)	Functions recover more slowly and more slowly after disturbance	'Critical Slowing Down' of recovery rates (i.e. engineering resilience) after small disturbance is an early warning signal of proximity to a tipping point (i.e. of low ecological resilience)	Distance to tipping point or regime shift	Cellular automata show that slowing recovery rates indicate proximity to thresholds for plants in semi-arid systems (Bailey, 2010; Dakos et al., 2011)	Veraart et al. (2012) show slowing recovery rates are a robust indicator of tipping point proximity (*cyanobacteria*); Dai et al. (2013) show 'recovery length' (the distance of recovery for spatially connected populations) increases with tipping point proximity (*yeast populations*)
Spatial or temporal autocorrelation (Scheffer et al., 2015)	System state variables become more similar to each other in space and/or time	Increasing autocorrelation is an early-warning signal of proximity to a tipping regime shift	Distance to tipping point or point	Autocorrelation does not always increase near a critical transition (Dakos et al., 2015) and is affected by spatially heterogeneous stressors (Génin et al., 2018)	Good indicator of critical transition in some systems (e.g. Veraart et al., 2012) (*yeast populations*) but not others (e.g. Bestelmeyer et al., 2013) (*plant communities*)
Spatial or temporal variance (Carpenter and Brock, 2006; Thrush et al., 2009)	The variability of functioning over time or space	Increase in function variance is an early-warning signal of proximity to a tipping regime shift	Distance to tipping point or point	Temporal variance not a robust indicator (Dakos et al., 2012); spatial variance may be unreliable (Génin et al., 2018)	Temporal variance does not relate to threshold proximity (Veraart et al., 2012) (*yeast populations*), but spatial variance was a good indicator (Eby et al. 2017) (*plant communities*)
Self-regulation (Cabell and Oelofse, 2012)	Degree to which a function can maintain itself	Feedbacks such as density-dependence make the function responsive and able to adjust to changes	*Recovery* by being able to adjust to variable conditions	Self-regulation is positively related to function persistence and ecological integrity (Mora, 2017)	Self-regulation by ecosystem engineers benefits persistence of soil fertility (McKey et al., 2010)* (*soil fertility*)
Exposure to disturbance (Cabell and Oelofse, 2012)	Function is exposed to frequent, low-intensity disturbances	Small disturbances foster the ability of function providers to maintain the function under further disturbance	*Resistance and recovery* through adaptation to disturbance		Microbial organisms that survive after freezing can adapt to changed temperature conditions (Stres et al., 2010) (*microbial respiration*)
Local coupling (Cabell and Oelofse, 2012)	Function relies on local resources/interactions (few imports) and produces little waste (few exports)	Functions are supported by species and resources that are adapted to the (historic) range of local disturbances and do not depend on external inputs/outputs	*Resistance and recovery* through local adaptation to disturbance		Vanilla production outside its native range relies on (human) hand pollination due to the absence of native pollinators (Garibaldi et al., 2009)* (*pollination*)

*Evidence from biodiversity-driven functions in agroecosystems (as opposed to evidence from biodiversity-driven functions in other systems).

similar responses to disturbances. In addition, the loss of species is non-random when associated with a particular disturbance, thus species with similar responses may be affected disproportionately, rather than species with different responses being affected equally by disturbance (Elmqvist et al., 2003; Mori et al., 2013). Further, classic metrics of taxonomic diversity provide little to no information about probable mechanistic links between species richness and function resilience, as they lack information about the relative importance and interactions of individual species and thus treat every species as equally important for the provision of a function (Cadotte et al., 2011; Gagic et al., 2015; McGill et al., 2006). Yet, the importance of a species' contribution in comparison with others in a community depends on species-specific functional traits ('effect' traits) related to their impact on ecosystem function. Losing a species of high importance will have a stronger effect on the provision and resilience of a service than the loss of a less important species. Consequently, recent studies have progressed from examining taxonomic diversity, to assessing further indicators of function resilience including the presence of key species or functional groups, functional diversity, redundancy and response diversity.

3.1.2 Presence of key species or functional groups

The presence or absence of *key species or functional groups* providing a particular function may drive both provision and resilience of the function (i.e. the functional identity hypothesis; e.g. Grime, 1998). In this case, the loss of key functional groups in response to disturbance may lead to a collapse of functioning. Theoretically, the consequences of the number of species removed from a system have been shown to depend on the functions those species perform in the system (Dunne et al., 2002). Empirically, extinction or lack of specialist pollinators has led to a collapse of pollination services and costly re-organization through hand pollination to maintain any amount of production, as in the case of vanilla but also apple production in Southern and Central Asia (Anderson et al., 2011; Garibaldi et al., 2009; Partap and Ya, 2012). Other examples include the necessary presence of avian frugivores to maintain seed dispersal, as shown in a natural experiment where frugivores were extirpated on one of two islands (Caves et al., 2013); and the importance of an insectivorous bird species for persistence of biological pest control in cacao (Maas et al., 2015). In soils, resilience (persistence or engineering resilience) of disease suppression may be due to the effect of one or a few groups (species or isolates) of organisms that are antagonists to the pathogen or active in disease suppression in other ways (Weller et al., 2002).

3.1.3 Functional diversity

Functional diversity measures the number or diversity of functionally disparate species or properties of a community. It is assessed through a variety of indices, such as functional evenness and functional divergence that have been found to be better predictors of ecosystem functioning than species richness (Gagic et al., 2015). Several mechanisms are thought to underlie the importance of functional diversity for resilience of functioning: (1) High functional diversity may, through sampling effects, lead to the presence of functional groups with key roles for functioning (see above). (2) In agroecosystems, high functional diversity may enable communities providing a given function to be effective in a range of crops with different traits and management (i.e. the functional complementarity hypothesis; Díaz and Cabido, 2001; Gagic et al., 2015). Although direct 'trait-matching' of crops and service providers has been shown to be more important than functional trait diversity for current provision of pollination services (Garibaldi et al., 2015), high functional diversity may represent essential insurance for resilience of functions to changes in the portfolio of planted crops, and/or to increased diversity of planted crops (as promoted by the European CAP Greening pillar; European Commission, 2016). (3) Because high functional diversity is linked to high occupation of niche space, this indicator may imply buffering against invasions of alien species and/or the addition of new functional groups. For example, disease suppression in soil can be related to unspecific competition for niche space from soil microorganisms (Termorshuizen et al., 2006). Further, ecosystems with low functional diversity are generally considered more prone to invasion by new species than functionally rich ecosystems (Elton, 1958; Levine et al., 2004). In line with this, functionally diverse soil bacterial communities have been found to be more resistant to invasion by exotics, including plant pathogens (Eisenhauer et al., 2013; Irikiin et al., 2006; Matos et al., 2005; van Elsas et al., 2012; Wei et al., 2015), as has the productivity of functionally diverse grassland plant communities (Fargione and Tilman, 2005). These effects have been explained by a high number of species interactions and intense competition for niche space. However, further evidence for the importance of functional diversity and complementarity for resilience of functions is currently lacking.

3.1.4 Functional redundancy and response diversity

The combination of *functional redundancy* and *response diversity,* respectively, guarantees that many species are able to provide a given function, and that they respond in different ways to disturbance (Elmqvist et al., 2003;

Mori et al., 2013; Oliver et al., 2015a). Currently, these measures are considered major drivers of the resilience of ecosystem services to disturbance. In contrast to species complementarity and niche specialization (Poisot et al., 2013), *functional redundancy* increases with the level of similarity, or functional niche overlap, between species in a community (Fonseca and Ganade, 2001; Naeem, 1998; Pillar et al., 2013). Ecological theories predict that communities with high functional redundancy can reduce the impact of loss of species on service provision (portfolio effect) or of species that experience a population decline as a result of global change (insurance hypothesis) (Hooper et al., 2005; Thibaut and Connolly, 2013). In recent years, an increasing number of studies have investigated the importance of functional redundancy for the resilience or variability of functional groups and ecosystem services under disturbance and change. For instance, Hallett et al. (2017) showed that pollination by wasps can effectively compensate for experimental removal of bumblebees, and Sanders et al. (2018) found that low trophic redundancy can lead to cascades of secondary extinctions and decreased persistence of parasitism following the removal of single parasitoid species. For the functioning of soils, there is a consensus that 'general functions' based on the performance of generalist species (or genotypes), like organic matter decomposition, are more resistant to disturbances than specific functions based on the action of specialist species, like nitrification (Deng, 2012; Griffiths and Philippot, 2013). This difference can be attributed to the observation that general functions are maintained by a wide range of redundant microorganisms, while specific functions are maintained by specific groups including fewer redundant species (Griffiths et al., 2001).

However, functional redundancy in itself is insufficient to ensure the resilience of functioning. A diversity of responses to disturbance and environmental change within redundant functional groups is seen as an additional necessary component (Elmqvist et al., 2003). *Response diversity* of ecosystem function providers in relation to environmental conditions enables compensation and spatiotemporal complementarity between species and individuals providing a particular function (Mori et al., 2013). Response diversity ensures that species with the same or a similar set of functions together contribute to functioning by being able to perform under varying environmental conditions, thereby potentially decreasing the variability of the function over time and increasing its resilience to disturbances. Different responses and sensitivity of species, and of individuals within species, to environmental change can be driven by: variation in inter- and/or intraspecific genetic makeup, phenotypic plasticity, and intrinsic growth rates (Oliver et al., 2015a). In a flexible modelling framework, Norberg et al. (2001) demonstrated theoretically that response diversity increases the persistence of functioning under disturbance by enhancing its capacity for resistance. Empirically, field studies below-ground have shown that selected isolates of microorganisms, applied as augmentative biocontrol to soil or plants to control diseases, can be highly effective in species-poor environments (e.g. in the laboratory or in otherwise sterile soil) but less or ineffective under field conditions (Alabouvette et al., 2006a; Cook, 1993). One way to find strategies that are effective across a range of temperatures and soil types is to combine biocontrol agents with different environmental preferences (Elead et al., 1994; Guetsky et al., 2001). Above-ground, the importance of frugivore response diversity for persistence of seed dispersal has been demonstrated (Farwig et al., 2017; González-Varo et al., 2017; Moreira et al., 2017). However, empirical evidence of the effects of response diversity on the resilience of crop-associated functions such as pollination is still limited and ambiguous (Cariveau et al., 2013; Rader et al., 2013; Stavert et al., 2018). Importantly, the benefits of response diversity and functional redundancy for function resilience in crops may depend on the type of disturbance considered (Balvanera et al., 2006). In addition, detecting the importance of these indicators may hinge on considering the full pool of organisms providing a function irrespective of provenance or specialization (Stavert et al., 2018).

3.1.5 Cross-scale resilience, negative co-variation, and effect vs. response traits

Further biodiversity-based indicators of cross-scale resilience, negative co-variation and the *relationship between species' response and effect traits*, are closely related to the above concepts of functional redundancy and response diversity. *Cross-scale resilience* is obtained when species or functional groups respond to disturbances at different spatial or temporal scales (Angeler and Allen, 2016; Elmqvist et al., 2003; Peterson et al., 1998). Similarly to the mechanisms of response diversity, cross-scale resilience implies risk-spreading of disturbance effects across scales (instead of between species). This benefits resilience by increasing functions' resistance to disturbances. In addition, due to nestedness of scales within each other, species only affected by disturbance at one scale can be recruited from other scales if they are mobile enough (e.g. landscape or regional species pools) and can thus perform functions at the disturbed scale, leading to resilience through recovery of functioning. Species able to enact such transfers can be considered mobile link organisms (see further definition below) contributing to spatial resilience (Allen et al., 2016). The effectiveness of function recovery across scales may depend on relative patterns of alpha and beta-diversity of functionally redundant mobile link species, e.g., consumers (Tscharntke et al., 2008a, 2012). However, few studies to date have examined the links between cross-scale diversity patterns in agroecosystems and function resilience. One study examining variability of wasp and bee parasitism over time (Tylianakis et al., 2006) found that temporal variation in parasitoid alpha diversity, but not high beta-diversity, led to less variable parasitism. However, positive effects of different responses of organisms between scales have been found outside agroecosystems, for instance on the persistence of avian spruce budworm predation (Holling, 1988) and recently for the recovery of coral reefs (Nash et al., 2016). In both examples, body size of predators was used as a proxy to infer differences in scales of responses of organisms to disturbance. In agroecosystems, data on body size of function providers is available for many taxa (e.g. carabids, staphylinids, spiders, bees, butterflies and moths; Boetzl et al., 2018; Duflot et al., 2014; Gossner et al., 2015; Öckinger et al., 2010; Williams et al., 2010). Approaches for cross-scale comparisons of (functional) diversity patterns are also well developed (e.g. Martin et al., 2016; Steffan-Dewenter et al., 2002). However, studies that explicitly link scale-dependent species' responses to the resilience of agroecosystem functions are lacking. In this context, research testing the 'intermediate landscape hypothesis' (Tscharntke et al., 2012) may represent a key step forward by showing that sufficient species pools at the landscape scale can compensate for disturbances or lack of resources at a local scale (e.g. Concepción et al., 2012; Jonsson et al., 2015).

Negative co-variation occurs when the abundance of pairs of species providing the same function is negatively correlated, due to either competition or variation in their response to environmental factors (Elmqvist et al., 2003). Similarly to response diversity, this indicator should lead to persistence of functioning through the continued performance of different species under disturbance (Hooper et al., 2005). Negative co-variation has rarely been found to occur in real communities as shown in several long term and/or cross-taxonomic studies (Houlahan et al., 2007; Karp et al., 2011; Valone and Barber, 2008), and thus appears not to be a particularly useful indicator of the resilience of functioning. However, for example, if there is strong competition between exotic and native species, such patterns may nevertheless occur and lead, counter-intuitively, to exotic species contributing to enhanced resilience (Stavert et al., 2018).

In contrast, the nature of the *relationship between effect traits and response traits* of organisms has been introduced and is gaining momentum as a likely indicator of resilience (Díaz et al., 2013; Oliver et al., 2015a; Standish et al., 2014; Suding et al., 2008). Effect traits are traits of organisms that determine their effects on a given function (e.g. consumption rates for predators). Response traits determine the response of organisms to environmental factors and disturbances (e.g. dispersal ability for arthropods; Lavorel and Garnier, 2002; Suding et al., 2008). If response and effect traits of organisms are correlated, then the loss of a response group after a disturbance implies the loss of the corresponding effect on function, even if other responses occur in the community. By contrast uncorrelated response and effect traits imply a balanced distribution of effect traits among responses of the community, and thus a decoupling between the vulnerability of functioning and the loss of particular response groups (Oliver et al., 2015a; Standish et al., 2014; Suding et al., 2008). Although still rarely examined in agroecosystems or explicitly linked to the persistence of agroecosystem functions (Díaz et al., 2013; Suding et al., 2008), correlated response and effect traits have been shown to cause a disruption of functioning under disturbance in the cases of both pollination and dung burial (Larsen et al., 2005).

3.1.6 Mobile links and biological legacies

The presence, redundancy and response diversity of *mobile link organisms* and *biological legacies* is seen as a crucial element for resilience of biodiversity-driven functions in agroecosystems, due to the spatiotemporal patchiness of these systems in terms of both structure and disturbance patterns (Folke et al., 2004; Lundberg and Moberg, 2003). Mobile link organisms represent species or individuals able to recolonize patches after a disturbance. Biological legacies, in contrast, remain in disturbed patches and form sources of regrowth. Because mobile link organisms effectively drive the transfer and recovery of functions through space, they are likely to constitute the key biological mechanism underlying why structural measures of 'spatial resilience' (sensu Allen et al., 2016), such as landscape heterogeneity or autocorrelation (the degree of aggregation of landscape patches), can influence the resilience of biodiversity-driven functions. Indeed, agricultural landscape heterogeneity represents a key factor of species' mobility (Fahrig et al., 2011; Schellhorn et al., 2014). Although many mobile link species are essential providers of valued services (see Section 3.1.2), others such as mobile pests or invasive species often promote undesirable agroecosystem states and 'disservices'

(Lundberg and Moberg, 2003; Standish et al., 2014). This is in spite of disturbances that aim at their elimination such as the use of pesticides against agricultural pests (Krauss et al., 2011). Particularly above-ground, the provision and resilience of pollination and biological pest control are contingent on the ability of pollinators and natural enemies to recolonize fields at appropriate phenological stages after planting, overwintering, or punctual destructive treatments during the growth season (pesticide application, tillage, mowing) (Schellhorn et al., 2015; Tscharntke et al., 2012). The mobility and dispersal ability of organisms enable not only recolonization for recovery of functioning in fields after disturbance, but also the survival of organisms outside fields for the duration of the disturbance, provided appropriate 'refuge' habitats and/or resources are within reach in the surrounding landscape (Bianchi et al., 2006; Schellhorn et al., 2014, 2015).

3.1.7 Interaction network complexity

In addition to previous indicators, the *structure of interaction networks* is also considered to be important for the ability of communities to withstand disturbance. Particularly the degree of nestedness, modularity and connectance of networks are expected to play a role in the resilience of network functions (Biggs et al., 2015; Tylianakis and Morris, 2017). In soils, this is particularly relevant when faced with the invasion of species, including plant pathogens. Soil communities with low nestedness and high connectance have been shown to resist pathogen invasion better than communities with low connectance (Wei et al., 2015). This could be explained by a more efficient use of resources in highly connected microbial communities, leading to a more intense competition for resources and reduced saprotrophic growth of the invading pathogen (Case, 1990; Wei et al., 2015). Theoretically, resource use or trophic complementarity of communities has been shown to be an important driver of the persistence of functioning, e.g., in the case of plant productivity (Poisot et al., 2013). In modelled plant-pollinator networks, high connectance and/or nestedness are shown to increase the ability of communities to avoid collapse, persist and recover after disturbance (Lever et al., 2014; Thébault and Fontaine, 2010). However, implications for the resilience of the pollination function itself, particularly in crops, remain unclear (Tylianakis and Morris, 2017). These are likely to depend on the effectiveness of trait-matching between pollinators and crops (e.g. Fontaine et al., 2005), which has yet to be integrated into recent modelling approaches (Bartomeus et al., 2016; Garibaldi et al., 2015). Theoretical frameworks provided evidence that in antagonistic networks (e.g. predator-prey), in contrast to mutualistic ones, modularity and low connectance foster community persistence and recovery (Thébault and Fontaine, 2010). However, the consequences for persistence of functioning itself also remain unclear (Tylianakis and Morris, 2017). For example, Macfadyen et al. (2009) found no link between the trophic structure of herbivore-parasitoid networks and the persistence of natural pest control on farms. Instead, the presence of effective parasitoid species was a determining factor (see the 'functional identity hypothesis' above).

3.2 Other indicators of resilience

3.2.1 Early-warning signals

Considerable effort has been spent in recent years on understanding and identifying possible 'early-warning signals' of an (eco)system's proximity to critical thresholds or tipping points, which if passed may lead to shifts in the system's structure, function, identity, and feedbacks (Scheffer et al., 2015). Such signals have been termed indicators of Critical Slowing Down (Dakos et al., 2015), because ecosystems approaching their tipping points have been found to recover more slowly from disturbance (thus being less ecologically resilient) than ecosystems that are far from their tipping points (Dakos et al., 2015; Scheffer et al., 2015). Indicators of an ecosystem or ecosystem function slowing down before transitioning to a contrasting state include decreasing *recovery rates* after disturbance (reflecting engineering resilience; Scheffer et al., 2009), increasing *spatial or temporal autocorrelation* (Dakos et al., 2010), and increasing *spatial or temporal variance* (Carpenter and Brock, 2006).

Direct measures of slowing *recovery rates* after disturbance have proven to successfully predict the proximity of ecosystems and functions to tipping points, both theoretically (Bailey, 2010; Dakos et al., 2011) and empirically (Dai et al., 2013; Veraart et al., 2012), in studies from semi-arid grasslands (Bestelmeyer et al., 2013) to experimental microcosms (Veraart et al., 2012). In contrast, rising *spatial or temporal autocorrelation and variance* sometimes, but not always, predict the onset of tipping points (Carpenter and Brock, 2011; Dakos et al., 2012). In some systems these indicators may even decrease before critical transitions, causing several authors to emphasize the need for caution and good system knowledge before interpreting them (Dakos et al., 2015; Génin et al., 2018). In addition, most of the work evaluating the performance of early-warning indicators has not been performed in agroecosystems, where inherent variability and heterogeneity of cropping patterns may strongly impact the meaning and applicability of such signals

(Vandermeer, 2011). Instead, work has focussed on natural ecosystems such as lakes and semi-arid grasslands (Dakos et al., 2010). Recently, Génin et al. (2018) examined for the first time the behaviour of early-warning indicators in spatially structured ecosystems subject to spatially heterogeneous stress, as is the case in agroecosystems (mussel beds, dryland vegetation and forest). They show that heterogeneous stress can confound expected trends of the indicators based on patterns of critical slowing down and spatial patch structure (Génin et al., 2018). Thus, it remains to be determined to what extent signals of critical slowing down can be applied to infer the resilience of agroecosystem communities and functions across disturbance gradients (Dakos, 2018; Dakos et al., 2015).

3.2.2 Self-regulation, exposure to disturbance and local coupling

Further potential indicators of the resilience of agroecosystem functions include their *ability of self-regulation*, their *exposure to low levels of disturbance* and their *local coupling* (i.e. the degree to which functions depend on locally available, as opposed to externally sourced, resources and interactions). These indicators are identified as particularly relevant in agroecosystems (Cabell and Oelofse, 2012). But to date, tests of their association with the persistence, engineering or ecological resilience of biodiversity-driven functions are scarce. Instead, studies have mainly examined the impact of these indicators on the resilience of the agroecosystem as a holistic social-ecological system (sensu e.g. Folke et al., 2010) without assessing the resilience of specific functions (Cabell and Oelofse, 2012; Peterson et al., 2018). In theory, *self-regulation* and internal feedbacks such as density-dependent functioning are likely to be positively related to function persistence and ecological integrity (the ability of an ecosystem to support and maintain an adaptive biological system with the full range of elements and processes expected in the natural habitat of a region; Mora, 2017). In a specific example, self-regulation by ecosystem engineers (termites, earthworms, ants) of pre-Columbian 'raised-field' cultivation systems may have contributed to the long-term persistence of soil services and fertility by contributing to, e.g., nutrient cycling, soil structure improvement and organic matter content (McKey et al., 2010). Mild *exposure to disturbance* that does not push function providers beyond survival thresholds may foster their ability to adapt and recover from further disturbances through, e.g., phenotypic plasticity (Oliver et al., 2015a), which may increase their resilience to other drivers of change in the future (Kühsel and Blüthgen, 2015). Finally, *locally coupled* functions that rely on the biodiversity of locally available and/or native organisms, such as pollination of traditional crops by native pollinators (Partap and Ya, 2012) or local availability of parasitoids of pests, are likely to increase the persistence, engineering and ecological resilience of functions by preventing reliance on a small number of imported species with low local adaptation or with the potential for invasiveness and competition.

4. Resilience of agroecosystem functions to environmental change

In the previous section, we outlined existing indicators with the potential to assess the resilience of functioning in agroecosystems, and their degree of validation by theoretical and empirical studies. Here, we examine what lessons can be learned from the use of such indicators, as well as from measures of function resilience itself. Specifically, we ask: (i) *how resilient are agroecosystem functions to disturbances related to environmental change?* and (ii) *how is function resilience affected by disturbances related to environmental change?* To assess these questions we focus on three biodiversity-driven, regulatory ecosystem services: (1) biological pest control, (2) disease suppression in soils, and (3) crop pollination.

4.1 Biological pest control

The general term biological pest control encompasses both the effect of naturally occurring enemies and antagonists, and of introduced or augmented biological control agents that reduce populations of different pestiferous organisms (Eilenberg et al., 2001). Here we focus on naturally occurring predators, parasitoids and pathogens of arthropod pests (biological control of plant diseases is discussed in Section 4.2). Biological control can help to keep damaging effects of pest species on crops below economically significant thresholds and thus reduce the need for direct pest control measures such as insecticide application. A key element of the effectiveness of biological control of agricultural pests is the presence and maintenance of a high abundance and diversity of natural enemies (Jonsson et al., 2017; Landis et al., 2000). However, antagonistic interactions among enemies can influence their potential to effectively control pests. In addition, the fact that predation is often density dependent means that measures of enemy abundance and diversity are often insufficient to assess the potential of a community to provide effective biological control. As a result, biological control potential is typically assessed using predator exclusion cages with standardized densities of (pest) prey, or by estimating attack rates on sentinel prey (Birkhofer et al., 2017; Lövei and Ferrante, 2017).

The majority of studies investigating the effects of environmental change on biological pest control itself are of a snap-shot character, and are thus not able to tease apart the temporal responses of biological control to disturbances. These studies mainly examine how persistent biological control services are under varying strengths and types of environmental disturbance. Indirect evidence that disturbances affect engineering and ecological resilience of biological control thus far mainly stems from studies of biodiversity-based resilience indicators. Based on that, we review the consequences of disturbances for the persistence of this service and for indicators of its resilience based on characteristics of natural enemy communities separately.

4.1.1 Land use intensity at landscape and local scales

The effects of disturbances related to land use intensity on predator biodiversity and biological pest control have been extensively studied both at field and landscape scales, sometimes also considering interactive effects between scales. When agricultural intensification is realized at the landscape level, a significant loss and/or fragmentation of natural and semi-natural habitat is often the consequence. A number of studies have explored the persistence of biological control in relation to land use intensity gradients at the landscape level. Rusch et al. (2016) re-analysed 10 datasets assessing aphid control with predator exclusion cages, and found that there was a consistent reduction in pest control services with landscape simplification (46% lower pest control in landscapes dominated by cultivated arable land). However, in a recent global re-analysis of exclusion cage studies no consistent effect of landscape simplification was found across studies, with approximately equal proportions of studies showing positive, negative or no effects of landscape simplification (Karp et al., 2018). Thies et al. (2011) studied biological control of aphids in cereal crops across Europe and found that pest control by natural enemies was reduced by landscape simplification in some regions but not in others. Thus, it appears that biological control in certain crop-pest-natural enemy systems in specific regions is more persistent to disturbances associated with landscape effects of modern agriculture than in others.

Across taxa, habitat degradation in agricultural landscapes has been linked to losses in the diversity and abundance of natural enemies. For instance, a reduction in suitable habitat in agricultural landscapes had negative effects on parasitoid communities and parasitism rates (e.g. Kruess and Tscharntke, 1994, 2000; Menalled et al., 2003; Thies and Tscharntke, 1999). Habitat loss or degradation can also lead to a reduction in the diversity of birds and bats and limit their potential for pest control (e.g. Faria et al., 2006; Kalda et al., 2015; Perfecto et al., 2004; Redlich et al., 2018; Tscharntke et al., 2008b). Furthermore, this factor has been linked to a reduction in the biodiversity of ground-dwelling arthropods with negative effects on the pest control services they provide (Nurdiansyah et al., 2016; Rusch et al., 2013, 2016; Weibull et al., 2003; Woodcock et al., 2010).

Response diversity of natural enemies to land use intensity has been studied for both functional and taxonomic groups. For example, Martin et al. (2016) found that the response of natural enemies to landscape simplification differed between taxa, with five out of seven broad natural enemy taxa being negatively affected by a simplified landscape configuration, but only carabids being negatively affected by reduced amounts of semi-natural habitat in the landscape. Typically, individual species and functional groups of species contribute differently to biological control services and if dominating key species or groups are particularly sensitive to disturbance this could have strong implications for resilience (see Table 1). Martin et al. (2013) found that flying predators were the most functionally important group for biological control of lepidopteran pests on cabbage in South Korea, but its effects on biological control were also the most sensitive to landscape simplification. Maas et al. (2015) studied predation in Indonesian cacao plantations and found that certain insectivorous birds with key importance for biological control were particularly sensitive to the distance to natural forest.

At local scales, a range of studies have explored how species diversity and biological control are affected by organic farming by comparing fields or farms under organic and conventional (intensive) management. Tuck et al. (2014) reviewed such studies and showed that the diversity of predators and several other functional groups was higher on farms under organic management. Östman et al. (2003) used predator exclusion cages combined with modelling to show that biological control was about twice as high in organic compared to conventional barley fields in Sweden. Together, these studies indicate that (1) biocontrol is not persistent to conventional management and (2) if maintained, conventional management should lead to decreased resilience of biocontrol to further disturbances.

Other studies have focused on the effects of individual disturbances. In a large European study, insecticide application was found to consistently reduce both predator diversity and biological control services (whereas fertilization levels had no effect) (Geiger et al., 2010). Insecticides often increase mortality of the natural enemies, but there is also evidence that it may modify natural enemy behaviour. For example, Stapel et al. (2000) found that insecticide application in cotton fields temporarily affected the foraging ability of a parasitoid wasp, thus reducing its ability to successfully control pest species for up to 18 days. Insecticides are known to affect species of natural enemies differently (Pisa et al., 2015), depending on their physiology and exposure to insecticide applications. Thus, it is clear that there is a response diversity for natural enemies in relation to insecticides, as long as enemy communities include species or individuals with variable responses. Biological pest control provided by such communities can be expected to be resilient against insecticide application. However, the available evidence (Geiger et al., 2010; Östman et al., 2003) suggests that in conventional European crops, this mechanism is currently either lacking or insufficient to maintain persistent biological control under insecticide use. In addition, where a degree of control is maintained, lower diversity of enemy communities on conventional farms and under insecticide application is likely to imply lower resilience of biological control to further disturbances.

Herbicides rarely have direct lethal effects on arthropods, but application of certain types of herbicides such as glyphosates have been found to modify arthropod (predator) behaviour (Korenko et al., 2016). The main effect of herbicide application is, however, in most cases indirect, acting via reduced habitat heterogeneity and reduced availability of alternative food sources (Nyffeler et al., 1994). Accordingly, it has been shown that predator abundance and diversity are reduced both when herbicides are applied (Asteraki et al., 1992) and when weeds are manually removed (Diehl et al., 2012). In a recent study, Staudacher et al. (2018) found that herbicide application induced a rapid change in predator-prey network structure, with increased levels of prey specialization across generalist predator species, probably as a consequence of enhanced competition among predators (Staudacher et al., 2018). Thus without herbicide application, niche-overlap was larger among predators suggesting that the level of functional redundancy was higher.

Intensive soil tillage can also have negative effects on natural enemy biodiversity. Tamburini et al. (2016) showed that conventional tillage including ploughing had negative effects on both predator abundances and their potential to control pests, but only in simplified landscapes. This provides evidence that complex landscapes can provide cross-scale community resilience making biological control persistent to local field-level disturbances.

Thorbek and Bilde (2004) studied the effects of several agricultural management measures related to disturbance (ploughing, non-inversion tillage, superficial soil loosening, mechanical weed control and grass cutting) on abundances of ground dwelling predators. All measures were found to reduce predator abundances, and the predators seemed to aggregate in less disturbed areas.

In recent years, food-web approaches have been increasingly used to demonstrate community-wide effects of land use intensification on biological control services. In particular, habitat degradation has been demonstrated to change the complexity of interaction networks and to alter interaction strengths between providers of pest control services and their prey. Laliberté and Tylianakis (2010), for example, have demonstrated that deforestation in tropical agroforestry systems leads to a spatio-temporal simplification of parasitoid-host networks, resulting in a homogenized interaction composition, and thus reduced potential resilience across rice and pasture sites in comparison to forested habitats. Gagic et al. (2012) report overall lower parasitism rates despite a higher complexity of the food web structure in an aphid-parasitoid-hyperparasitoid system in areas of high agricultural intensification.

A few studies have assessed spatial or temporal variability in biological control services and related these to disturbances associated with land use intensity. Tylianakis et al. (2006) found lower variability of parasitism rates when parasitoid diversity was high, which happened in this case in highly modified habitats. Similarly, Macfadyen et al. (2011) found lower variability of parasitism rates when parasitoid species richness was high, but in this study this occurred in organic (less disturbed) farms as opposed to conventional ones. In a unique attempt to test ecological resilience, Macfadyen et al. (2011) also simulated parasitoid removal from food-webs to assess the robustness of pest control under scenarios of extinction, but did not find any difference between farming systems. Rusch et al. (2013) studied within field variability in biological control using exclusion cages and found that the within-field spatial variability in biological pest control services decreased with crop rotation intensity in the landscape, although variability in parasitism rates increased.

To conclude, many studies have explored how different types of disturbances associated with land use at field and landscape levels affect various biodiversity-based resilience indicators, and a growing number also explore effects on biological control services themselves, using exclusion cages or sentinel prey. Applying space-for-time substitution (De Palma et al., 2018),

these studies show that persistence of biological control in the face of disturbance caused by land use intensity varies between crop-pest systems and
regions. Also, the effects appear to depend on the actual disturbance, with
insecticide application, for example, showing consistent negative effects on
biological control, whereas landscape effects seem to vary more. However,
the reasons for different levels of persistence are rarely known—even if they
seem to correlate in several cases with levels of natural enemy diversity. The
great majority of available studies are of short-term snapshot character and
do not examine temporal aspects of resilience, such as the ability of biological
control to recover or persist over the long term. Instead, a few studies have considered changes in spatial or temporal variability in relation to disturbance. To
date, effects of land use intensity on engineering resilience are few and ecological resilience has almost not been studied at all (but see Macfadyen et al., 2011).

4.1.2 Climate change

Studies investigating the resilience of biological pest control to climate
change are much more limited than those investigating resilience to land
use intensity. In general, models predict that pest problems will increase with
climate change in large parts of the world, except in the lowland tropics
(Deutsch et al., 2018; e.g. Diffenbaugh et al., 2008; Maiorano et al.,
2014). A recent global study of maize, rice and wheat, for example, predicted
that crop losses due to pests would increase by 10–25% globally, with the
largest increases taking place in temperate areas where current yield levels
are highest (Deutsch et al., 2018). However, such models are usually based
on pest biology (e.g. metabolic rates and population growth; Deutsch et al.,
2018), and ignore the potential impact of natural enemies, which could
buffer against or enhance the predicted changes in crop losses depending
on how enemies are affected by climate change.

   Across terrestrial ecosystems, species diversity generally increases towards
the equator (Hillebrand, 2004) and a recent global assessment of predation
rates showed that predation rates by arthropods increase towards the equator
and at lower elevations (Roslin et al., 2017). This suggests that predators are
generally both more diverse and have a stronger top-down impact on herbivore populations at higher temperatures. It remains unclear, however,
whether predators will be able to track the changes in climate to maintain
these patterns under future conditions. Indeed, it has been predicted that
biological control may be reduced under climate change in specific systems.
For example, a modelling study on host-parasitoid food webs predicted that
increasing ambient temperature could lead to a reduction in biocontrol

services in systems where tolerance for higher temperatures is lower in parasitoids than their hosts (Furlong and Zalucki, 2017).

Up to now, there is little empirical evidence on how temperature affects the structure of predator communities and the implications for resilience of biological control. In a unique experimental study, Drieu and Rusch (2017) found that diverse predator communities shifted from being redundant to being complementary in their effects on leaf hopper pests on grape wines when moving from ambient temperatures to a +3 °C global change scenario. Essentially this provides evidence for response diversity of predators in relation to temperature. Thus such diverse predator communities are likely to provide resilient biological control under climate change.

Drought can have complex indirect effects on biological pest control. In a mesocosm study, Barton and Ives (2014) found that water stress in alfalfa led to reduced growth rates of pea aphids, which led to fewer ladybeetle predators and ultimately reduced predation rates on spotted aphids and a release of this aphid species from top-down control. Effects on biological control may become even more difficult to predict if drought is combined with elevated temperatures. Using a combination of field observations and laboratory mesocosm studies of cabbage aphids and their parasitoids, Romo and Tylianakis (2013) showed that the parasitoids had a better ability to control aphids under increased temperatures and drought when studied separately, but that the results were reversed when the two disturbances were combined. However, the role of biodiversity in providing resilience against drought has not been explicitly studied.

In conclusion, the currently small number of studies on the effects of climate change on predator diversity and resilience of biological control services does not allow reliable predictions about their resilience in future climate scenarios. In addition, the potential to generalize results of individual studies to form such predictions could be limited, because responses are likely to vary with respect to direct and indirect impacts of climate change on both enemies and pests as well as their host plants, and because interactive effects of different aspects of climate change and other disturbances may be common (Sasaki et al., 2015; Thomson et al., 2010).

4.2 Plant disease suppression in soil

Soil-borne plant diseases are important threats to agricultural crops, resulting in severe yield losses. These diseases are often difficult to control by conventional strategies such as chemical disease control or the use of resistant cultivars (Weller et al., 2002). Therefore, protection and management of naturally occurring disease suppression in soils constitute an interesting possibility for sustainable disease control in agricultural crops. The ability to suppress diseases is a quality of the soil that can be enhanced or destroyed by environmental factors connected to global change, including cultural practices and other disturbances (Larkin, 2015).

All soils have some capacity to suppress soil-borne plant pathogens and the diseases they cause. Soils with strong disease suppression were described early on as suppressive soils, in contrast to conducive soils, where disease occurs readily (Baker and Cook, 1974). However, the ability to suppress diseases is better described as a continuum from strong disease suppression to very limited suppression (Alabouvette et al., 2006b). The mechanisms of importance for suppression vary with the ecological strategy of the pathogen, and the combination of pathogen and host plant (the 'pathosystem'; Termorshuizen and Jeger, 2008). Thereby, there is also variation in the ability of the suppression to withstand disturbances. Typically, the mechanisms behind the suppression are not fully understood. Suppression can be due to biotic or abiotic factors or the combination of both, and involve both suppression of pathogen growth and of disease development (Kinkel et al., 2011). Cases where the suggested mechanisms are purely due to abiotic factors will not be covered here.

In contrast to above-ground pest control, the available literature on disease suppression in soil mainly focuses on the scale of plots to fields, though some studies have investigated interactions of general soil functions (e.g. as approximated by soil organic matter content) with landscape-scale simplification (Gagic et al., 2017). Climatic disturbances and change may theoretically have strong impacts on disease suppression in soil and its persistence (Stres et al., 2010). However, few studies have investigated these effects considering realistic ranges of temperature or precipitation change. Typically, disease suppression in soil is assessed by performing bioassays where an isolate of the pathogen is inoculated into the soil, and a susceptible plant is grown under temperature and humidity favourable for disease development. Specific mechanisms or processes of importance for suppression are also commonly studied on nutrient media or other laboratory conditions, for example, restriction of fungal spore germination (fungistasis) or inhibition of microbial growth (microbiostasis; Ho and Ko, 1982; Lockwood, 1977).

4.2.1 Types of disease suppression and their sensitivity to disturbances

Disease suppression is divided into general and specific suppression, often working together to form the suppressive ability of a specific soil (Weller et al., 2002). General suppression is related to unspecific competition for niche space by the total microbial biomass in soil and its activity (Termorshuizen et al., 2006). This type of suppression is widespread, occurs more or less in all non-sterile soils, and is often enhanced by the addition of organic material. Especially in the case of competition sensitive pathogens, this type of suppression can be of great importance, for example, in the suppression of corky root of tomato, caused by the fungus *Pyrenochaeta lycopersici* (Hasna et al., 2007). As general soil suppression is caused by the total activity of the microbial biomass rather than specific species, it is not particularly sensitive to disturbances as long as the microbial activity is maintained or increased. One example of compromised general suppression can be seen after the drastic disturbance caused by sterilization of greenhouse soils in tomato production. After soil sterilization, Fusarium crown and root rot caused by *Fusarium oxysporum* f. sp. *radicis-lycopersici* developed readily in an environment with few competitors (Rowe, 1978). However, after sterilization, all soil organisms with more or less specific interaction with the pathogen are similarly killed. Thus, deleterious effects of soil sterilization on persistence of soil disease suppression can also be due to loss of organisms with more specific interactions (specific suppression).

Specific suppression is, in contrast to general suppression, suggested to be due to the effect of one or a few groups (species or isolates) of organisms that are antagonists to the pathogen or active in disease suppression in other ways (Weller et al., 2002). In some cases, specific suppression can even be due to the production of specific antagonistic substances such as antibiotics (Weller et al., 2002). The effect of a disturbance on specific suppression depends completely on how sensitive the important organism(s) are to the disturbance and how well they can recover from it. In several of the best studied examples of specific suppression, the organisms of importance have been isolated and transferred to other soils, resulting in disease suppression also in these soils (Weller et al., 2002). This transferability has been used to prove the role of a specific isolate in disease suppression, but requires that the organism that is isolated is competitive enough to colonize the soil to which it is transferred.

4.2.2 Field-scale disturbances: Agricultural management and crop rotation

For several soil-borne plant pathogens, soils with strong disease suppression have been identified. In these soils, the disease development is limited even in presence of pathogen inoculum and under environmental conditions favourable for disease development. Although the important mechanisms occurring in theses soils are not fully understood and are known to vary with the pathosystem, the suppressive activity is suggested to be complex in its nature, and caused by a combination of general and specific suppression (Schlatter et al., 2017; Weller et al., 2002). In certain cases, the disease suppression is long-standing, with little variation over years and not drastically changed depending on the crop rotation or cultural practices used. One example of such long-standing suppression is found in Fusarium wilt-suppressive soils of Châteaurenard in France, which has been attributed to the combined activity of certain non-pathogenic strains of *F. oxysporum* and the bacterium *Pseudomonas fluorescence* (Siegel-Hertz et al., 2018; Trivedi et al., 2017). The reason for persistence of the disease suppressing activity of this soil is not well understood. It has been suggested that the fact that the suppression is long-lasting serves as a proof for direct or indirect influence of physicochemical soil characteristics. The importance of environmental factors in influencing the structure and function of soil microbial communities is supported by Griffiths and Philippot (2013), Banning and Murphy (2008) and Orwin et al. (2006). According to these authors, the persistence in soil functioning over time, despite varying levels of disturbance, is a result of interactions between the microbial community structure and environmental factors such as soil type or nutrient availability.

Soil suppressiveness can also be induced, for example, by cultural practices like additions of organic material, specific crop rotations or by crop monoculture (Raaijmakers et al., 2009). For example, suppression of Fusarium wilt can be induced by continuous cropping of partially resistant cultivars. This induced suppression is suggested to be largely caused by non-pathogenic isolates of *Fusarium oxysporum*, and to be mainly due to induced pathogen resistance in the host plants. This is in contrast to the long-standing suppression in Châteaurenard described above, which also involves suppression of the pathogen's saprotrophic growth and spore germination (Larkin, 1996; Weller et al., 2002). Another example of monoculture-induced suppressiveness is take-all decline—a reduction of take-all of wheat (caused by *Gaeumannomyces graminis* var. *tritici*) after continuous cropping of susceptible cultivars and at least one severe outbreak of take-all. As a response to the build-up of pathogenic populations, a range of antagonistic microorganisms accumulate in the soil, resulting in strong disease suppression. This phenomenon has been described to occur worldwide, with some variation in its extent and speed of development. However, it is not long-standing and is lost when the monoculture is broken (Weller et al., 2002). In these examples, intensification of cropping practices illustrated by increased use of monocultures is thus shown to have a positive impact on the persistence of disease suppression in soil. This is in stark contrast to the effects of this driver generally observed on, e.g., above-ground pest control, and to findings that species-rich plant communities can lead to suppression of plant pathogens (Mommer et al., 2018) or increase the disease suppressive potential in soil (Latz et al., 2016).

4.2.3 Field to microcosm scale: Incorporation of organic material

Soil suppressiveness can be induced by addition of various types of organic material to the soil, or by leaving crop residues in the field to promote decomposers living on the residues (Raaijmakers et al., 2009). Addition and incorporation of organic material can be regarded as a type of disturbance of the soil system, changing the habitat for organisms living there, although in this case, the purpose is to increase the activity of beneficial organisms suppressing plant diseases. As such, this practice represents the inverse trend to depletions of soil organic matter characteristic of current agricultural intensification. The use of composted plant material to induce soil suppressiveness has been studied extensively and found to have the potential to suppress various types of plant diseases, through competition-based suppression as well as other mechanisms (Termorshuizen et al., 2006). The addition of compost material changes physicochemical and biological properties of the soil, and the microbiological changes are considered to be of particular importance. Effects of soil treatments also depend on site-specific soil characteristics. Pérez-Piqueres et al. (2006) studied the effect of four composts in two soils with differences in biological and physicochemical characteristics. In both soils, all four composts caused changes in bacterial and fungal community structures, and effects on community-level physiological profiles and suppression of damping of pine (*Pinus nigra*), caused by *Rhizoctonia solani*, varied depending on the combination of compost and soil.

Another type of plant material used to induce soil suppressiveness is that of Brassicas or other crops containing glucosinolates, a strategy called biofumigation. When such plant material is incorporated into the soil, the glucosinolates are hydrolysed into a range of products with a broad biocidal activity. This results both in direct reduction of microbial populations, including pathogens, and in secondary effects when saprotrophic organisms use the plant material as a food source. The relative importance of these two effects is debated, and depends on the chemical properties of the plant material and site-specific soil characteristics. The suppression has been corre-lated with both reductions in pathogen populations (Smolinska et al., 2003), with increases in microbial biomass or activity (Yulianti et al., 2007), with changes in the microbial community structure (Wang et al., 2014) or with increases in specific pathogen antagonistic populations (Mazzola et al., 2007; Weerakoon et al., 2012).

In biofumigation effects as well as effects of compost addition or other types of soil amendments, the effect of the disturbances on soil microbial communities and their ability to suppress diseases depend on the type and amount of material applied (Yohalem and Passey, 2011). In biofumigation treatments, the toxic products do not have any known long-lasting effect. Their main activity lasts from hours up to a few days (Matthiessen and Kirkegaard, 2006). Effects of the organic material itself last longer, as well as secondary effects from changes in the abundance and structure of com-munities. Many responses of the microbial community are seen directly after a biofumigation treatment, after which the community slowly returns to its initial biomass and community structure. Especially the bacterial community seems to be influenced by biofumigation treatments, but it appears to return to initial structure and biomass within a few months. In contrast, fungal communities are more resistant and take a longer time to return to initial structures (Friberg et al., 2009; Mocali et al., 2015; Wei et al., 2016). There is limited information about how the duration of responses in microbial communities relates to the suppression of diseases. In some cases, the biofumigation effect on disease suppression lasted longer than changes in microbial community structure were detected (Friberg et al., 2009). This indicates that the engineering resilience of the community to biofumigation may be higher than the resilience of functioning itself. At least in part, such lack of correlation in changes in population structures compared to the suppression could be due to differences in the sensitivity of methods used to characterize communities and assess suppressiveness.

To conclude, there are examples where disease suppression in soil is persistent to disturbances connected to cropping practices occurring over both long and short time scales. These include examples of long-lasting specific disease suppression resilient to variation in cropping practices and crops over time, but also examples where the suppression is highly depen-dent on specific practices and collapses if these practices change. Several studies have focused on the beneficial effects on suppressiveness obtained by adding or preserving organic material in the soil. Based on this, agricul-tural intensification leading to a depletion in soil organic matter should be considered as one of the most problematic consequences of global environmental changes. To date, resilience of disease suppression in soil has been addressed in terms of direct measures of function persistence and engineering resilience, including long term studies. In contrast, effects of disturbances on resilience indicators, including measures of community structure for disease suppressing organisms, are less frequently studied in this framework. Mechanisms of disease suppression are often complex and vary between sites and pathosystems. Through a better understanding of the processes and organisms of importance, agricultural practices and measures could be adjusted to, at least partly, counteract negative effects of global change on plant health.

4.3 Crop pollination

The evidence for a decline of bees (and more generally pollinators, Biesmeijer et al., 2006) raises alarm for the resilience of their supporting pollination services to pollinator-dependent crops (Potts et al., 2016). Indeed, less diverse pollinator communities may provide ecosystem services that are less resilient to disturbance over space or time (e.g. Rader et al., 2013; see discussion in previous section). Bees are the most important group of pollinators in temperate climates (Ollerton et al., 2011), and provide critical pollination services for wild plant communities and agricultural productivity (Potts et al., 2016). Nearly 90% of the world's flowering plant species (angiosperms) are pollinated by animal mediation (Ollerton et al., 2011), including ca. 70% of world crops (Klein et al., 2007). However, bees as well as pollinators in general are currently declining worldwide (Cameron et al., 2011; Goulson et al., 2015; Potts et al., 2010, 2016), and plant-pollinator networks are disturbed (Biesmeijer et al., 2006; Fitter and Fitter, 2002). For instance, during the last decades, the diversity of bees observed in Great Britain and the Netherlands has decreased by more than half, in parallel with the decrease of the diversity of plants (Biesmeijer et al., 2006). Crop pollination is therefore commonly cited as an example of an endangered ecosystem service and several studies have analysed the risk of pollination deficits, and their relations to global environmental changes that affect pollinator biodiversity.

4.3.1 Assessing crop pollination service

Assessment of pollination services in crops has been approached by different manners and metrics. The simple measure of the abundance and diversity of flying bees captured inside the crop has been used as a proxy of the visitation rate of flowers (e.g. Carvell et al., 2007; Le Féon et al., 2010, 2013). Pan traps are widely used for this purpose (but also transects with net catch). When placed in the crops, pan traps mimic flowers and collect visiting pollinators. This estimate is very approximate and potentially not related to actual pollination success, because collected bees may be either flower visitors or just crossing the field without visiting flowers. Moreover, no estimate of pollen deposition is available through this method. Other studies use the direct measure of the number of pollinator visits per flower as a proxy of pollination service (e.g. Bartomeus et al., 2014). This method allows to estimate the abundance and diversity of flower visitors. Nevertheless, the relationship between bee visit and pollen deposition is missing (i.e. the transport of pollen from the anthers [male organ] to the stigma [female organ]). This prevents any robust estimation of the pollination service itself as, for instance, some bees are known to 'rob' the nectar of plants without any process of pollen deposition (Saez et al., 2017). The most rigorous approach to estimate the pollination service of bees is to measure pollen deposition per bee visit. Some studies have counted the number of pollen grains deposited by bees on the plant stigma, after excluding the relative contributions of wind and self-pollination through exclosures (Cariveau et al., 2013; Kremen et al., 2002). In parallel, bee visits were also surveyed (i.e. the abundance and diversity of bee visitors was recorded). This rigorous estimate of pollination service is time-consuming (microscopic counting of pollen grains deposited on the stigma) and complex to replicate, which is why few studies have applied this method. A more frequently used alternative is counting the number of seeds produced per flower or plant (i.e. seed set), following the same protocol of exclusion (e.g. Garratt et al., 2014; Holzschuh et al., 2012; Porcel et al., 2018).

4.3.2 Impacts of land use intensity at landscape and local scales

We provide a synthesis of the principal effects of environmental change-related disturbances on the resilience of pollination services and its indicators in agroecosystems. We focus especially on bees due to their critical importance as pollinators, but studies with comparable metrics and patterns have considered other pollinator groups (e.g. Rader et al., 2016).

Agricultural intensification, i.e., land use change and habitat degradation, is considered a major cause in the decline of bee-driven pollination services in agroecosystems (Goulson et al., 2015; Potts et al., 2010, 2016). At the landscape scale, the loss and degradation of semi-natural habitats have reduced the amount and diversity of floral resources (Goulson et al., 2008; Williams and Osborne, 2009) and the availability of nesting sites for pollinators (Steffan-Dewenter and Schiele, 2008), resulting in a general decline in bee abundance and diversity in agricultural areas. Compelling evidence of positive effects of the proximity and amount of natural (or semi-natural) habitats on the abundance and diversity of bee visits in crops is available from several studies and syntheses (e.g. Cariveau et al., 2013; Kennedy et al., 2013; Kremen et al., 2002; Ricketts et al., 2008). These studies confirm the benefits of natural habitat for bee diversity in pollinator-dependent crops. However, the associated improvement of pollination service is undemonstrated. While some studies show an increase in pollen deposition with proximity to natural habitat (Cariveau et al., 2013; Kremen et al., 2002), others show no effect on fruit and seed set (Ricketts et al., 2008), thus implying persistence of pollination services under landscape simplification and habitat loss. Finally, some studies examined the engineering resilience (recovery) of pollinator communities after restoration or planting of semi-natural habitats, i.e., under a reversal of the trends in agricultural landscape simplification. These studies focus on bee abundance and diversity in crops (e.g. using pan traps) and not on pollination itself. They provide evidence of positive effects of wildflower planting and restoration of native plant hedgerows on the diversity of pollinator communities (e.g. Kremen and M'Gonigle, 2015; Williams et al., 2015), suggesting beneficial effects of these measures on the resilience of pollination to future environmental disturbance.

At a local scale, the use of agrochemical inputs is found to affect bees directly, e.g., insecticides induce sub-lethal effects on bee behaviour and survival (Henry et al., 2012). Indirectly, e.g., herbicides decrease the availability of floral resources and fertilizers decrease the diversity of in-field plants (Gabriel and Tscharntke, 2007; Goulson et al., 2015; Holzschuh et al., 2007). Recently, evidence has been found for negative effects of exposure to, e.g., neonicotinoid insecticides on wild bee density and flower visits (Rundlöf et al., 2015). In a uniquely long-term dataset monitoring bees in Great Britain over 18 years, these deleterious effects have been shown to extend to whole communities of bee pollinators, affecting both the persistence and likelihood of extinction of many species (Woodcock et al., 2016). In contrast, organic farming (without insecticide exposure) has been shown to increase the diversity and abundance of native bees in agroecosystems compared to conventional management (Kremen et al., 2002). However, expected stronger provision of associated pollination services (fruit set or pollen deposition) in organic management, which would imply low persistence of pollination under insecticide use, is controversial (see Kremen et al., 2002; Porcel et al., 2018).

4.3.3 Supplementation of managed pollinators, and other disturbances

Environmental change can negatively affect the biodiversity of bees and other pollinators (Biesmeijer et al., 2006) until a tipping point of pollination resilience is reached that could be viewed as either (1) the complete extinction of wild bees, or (2) the absence of trait-matching between local bees and crop flowers (e.g. Bartomeus et al., 2016; Garibaldi et al., 2015). The first (extreme) scenario occurs in some parts of Asia. In recent years, farmers have been forced to hand-pollinate apple trees, carrying pots of pollen and paintbrushes with which to individually pollinate every flower, after the extinction of local pollinators due to habitat degradation (Partap and Ya, 2012). The second scenario could occur with the current decline in bee diversity (see below; Biesmeijer et al., 2006; Potts et al., 2010, 2016). Supplementation of managed generalist pollinators is now common in agroecosystems, and is suspected to counteract gaps in wild crop pollination services (Garibaldi et al., 2017). For this purpose, the western honey bee (*Apis mellifera* L.) is the species that is most widely used across the world, although recent studies proposed management of other species (reviewed in Garibaldi et al., 2017; Isaacs et al., 2017; Pitts-Singer et al., 2018). The currently most common management practice to reduce potential pollination deficits in pollinator-dependent crops consists of increasing the stock rate of managed honey bee colonies per unit area (Isaacs et al., 2017).

However, artificial supplementation can have detrimental effects on wild pollinators, such as decreasing their flower visitation (by competition), reproductive success, abundance, and diversity (Elbgami et al., 2014; Geldmann and González-Varo, 2018; Geslin et al., 2017; Goulson and Sparrow, 2009; Hudewenz and Klein, 2013). Thus, artificial supplementation of pollinators can have detrimental effects similar to the introduction of invasive species. In this way, some managed pollinators that have been introduced for crop pollination out of their native range are currently invasive. In particular, the introduction of *Bombus terrestris* for the pollination of tomato in Chile has led to a large scale invasion throughout Latin America, and the collapse of native bumble bees through competition for resources (Morales et al., 2013). Thus, supplementation of managed pollinators can affect the resilience of the pollination service in agroecosystems. In addition, other threats may affect pollination service resilience such as the spread of invasive parasites and pathogens of pollinators, cross-transmission between managed populations and wild species, and impacts of climatic disturbances and change on shifts in the range of native pollinator populations (reviewed in Potts et al., 2010, 2016). However, to our knowledge, none of these studies have investigated these effects on the resilience of pollination services or associated indicators.

4.3.4 Resilience of crop pollination to disturbance: Major knowledge gaps

Recently, considerable progress towards measuring and predicting ecological resilience of pollination has been made in natural ecosystems (Fontaine et al., 2005; Jiang et al., 2018; Thébault and Fontaine, 2010). But in agroecosystems, the studies reviewed above show that direct resilience assessments relate especially to pollination persistence and to some extent engineering resilience (recovery) of pollinator communities. In addition, the majority of studies focus on the resilience indicators of species richness or diversity of bees, but few explore effects of disturbances on other indicators involving functional traits. Based on available studies, the resilience of pollination services in agroecosystems is likely to be threatened by multiple disturbances including habitat loss, pesticide use, supplementation of managed pollinators and climate change (Potts et al., 2016). Persistence of these services may be supported by organic practices and the restoration of seminatural habitats, but little is known to date about long-term effects of other environmental measures such as flower planting. Furthermore, a more realistic context of trait-matching of mutualistic interactions, as already described in natural ecosystems (Fontaine et al., 2005; Thébault and Fontaine, 2010) currently needs to be considered.

Trait-matching is the process by which pollinators have coevolved specialized mutualisms with flowering plants. It is characterized for instance by pollinators with long tongues mainly visiting plants with deep corolla, suggesting a strong match between flower and pollinator morphology at the individual scale (Bartomeus et al., 2016; Garibaldi et al., 2015). Due to the spatiotemporal heterogeneity of cropping patterns, extension of methods developed outside agroecosystems to assess ecological resilience of pollination in agroecosystems may be contingent on the consideration of trait-matching between pollinators and crops. Indeed, in a first step considering the provision (not the resilience) of pollination services, trait-matching has been shown to better predict success of crop fruit set than trait diversity (Garibaldi et al., 2015).

However, to our knowledge, no studies have analysed the effect of environmental disturbances on the persistence or recovery of trait-matching in agroecosystems. We hypothesize that such approaches would be robust measures of the resilience of pollination services. Moreover, to our knowledge no study has considered the implications of the planting of exotic crops for the resilience of pollination (or other services) in agroecosystems. Indeed, movements of wild bees from local (semi) natural habitats into crop fields are often expected to benefit the provision and persistence of pollination services in crops. However, this expectation does not consider the fact that most pollinator-dependent crops are exotic in the system. Pollinator-dependent crops are indeed frequently established whose flower traits do not match the traits of the local (native) bee community. Thus, (1) no local bee will be able to pollinate the crop, (2) the crop will reduce the availability of nesting and feeding semi-natural habitats for local bees, (3) farmers will need to practice supplementation of managed generalist pollinators (or 'human pollinators'; Partap and Ya, 2012), that can spillover into semi-natural habitat after the flowering period of the crop and therefore compete with native wild bees, reducing their fitness and affecting the resilience of the pollination service through a number of mechanisms (Table 1). This drastic scenario does indeed apply in the case of the use of plants with deep corollas (such as perennial leguminous herbs) that often require trait-matching interactions with very long-tongued bees, which are currently endangered (Cameron et al., 2011). Overlooking the trait-matching mutualistic interactions of pollination services should annul efforts of currently applied pollinator-friendly schemes (e.g. reducing pesticide exposures, increasing natural nesting and flowering habitats). Thus, we hypothesize that cultivating crop plants with trait-matching to local wild bee populations would enhance the resilience of pollination services against present and future environmental disturbances.

5. Operationalizing resilience: A critique and steps ahead

In the previous section, we reviewed to what extent measures of resilience to environmental disturbances have been operationalized to date for three key biodiversity-driven services in agroecosystems. We conclude that in contrast to certain other functions or services (e.g. biomass production; Isbell et al., 2015) and certain other ecosystems (e.g. coral reefs; Nash et al., 2016), measures of resilience for these functions lag behind in their ability to uncover mechanisms and to predict response trajectories to various types of disturbances. Indeed, for the functions considered, only very few studies have investigated even the more 'accessible' measures of engineering resilience (Ingrisch and Bahn, 2018) by testing recovery rates after disturbance. More complex measures of ecological resilience remain, for now, essentially uncharted territory (but see Macfadyen et al., 2009, 2011), despite developing approaches outside agroecosystems that may represent important stepping stones (e.g. for pollinator networks; Jiang et al., 2018).

Instead, many studies measuring the functions of interest focus on (1) the persistence of functions under disturbance or (2) on indicators (particularly biodiversity-based) of the resilience of functions. Generally, these studies are not framed to examine a component of resilience.

(1) In the case of measures of function persistence, studies on pollination and biological pest control most often use snap-shot measures that inform on the state of the function after disturbance. However, they provide no indication of whether or not the function may recover in the future, whether it is already on a trajectory for recovery, or may further degrade. As such, the conclusions that can be drawn for resilience over the longer term are limited. However, these studies nevertheless inform on the important aspect of to what extent a function was able to persist under disturbance, and thus—if measured in a standard way across systems—can help rank the resilience of functions in different agroecosystems subject to similar pressures (e.g. Cariveau et al., 2013; see also metrics of Category I, Ingrisch and Bahn, 2018). In this regard, several studies on plant disease suppression in soil are the exception by including longer term measurements of persistence and recovery. In addition, studies examining temporal variation of pollination and biological pest control after disturbance are still rare, but represent an important step in this direction (e.g. Macfadyen et al., 2011). These studies have yet to show their full potential in terms of resilience assessment by explicitly comparing trajectories of long-term persistence to undisturbed (dynamic) baselines (Egli et al., 2018; Ingrisch and Bahn, 2018).

(2) Particularly for functions above-ground, many measures of how communities, functional community structure and interactions are modified under different types of disturbance have been examined in agroecosystems. In the case of pollination, measures based on biodiversity appear to dominate even more than in the case of biological pest control compared to direct measures of persistence of the functions themselves. However, this trend could change due to the current development of trait-matching approaches in pollination studies. While trait-matching approaches are gaining ground in the context of pollination, they have yet to be developed in studies of biological pest control, despite strong benefits that can be expected in this field. In contrast to these functions above-ground, studies of disease control

in soil rarely examine biodiversity-based indicators compared to direct measures of the function itself. Although the validity of biodiversity-based indicators is often still far from demonstrated (Egli et al., 2018; Section 3; Table 1), if confirmed these measures could give key indications of the resilience of functions to future disturbances (e.g. Oliver et al., 2015a; Scheffer et al., 2009).

5.1 Conceptual hurdles and technical challenges

The distinction between the two types of measures—direct function persistence (1) vs. biodiversity-based or other resilience indicators (2)—in fact represents far more than a difference in methods. In Figs 3 and 4, we highlight the essential differences in the questions they can each address. In fact, when addressing the resilience of (agro)ecosystem functions under EC, two superficially similar but fundamentally different questions can be examined:

(i) To what extent are ecosystem functions able to persist (recover, resist) under EC? (Fig. 3)

(ii) What is the impact of EC on the ability of ecosystem functions to persist (recover, resist) under further disturbance? (Fig. 4)

We hypothesize that a lack of clarity about which of these questions is being addressed compounds the profusion of resilience metrics, definitions and theoretical assumptions (Donohue et al., 2016; Egli et al., 2018; Weise et al., 2019), hindering the development of operational frameworks in agroecosystems. In question (i), direct assessment of the function's response to disturbance is both necessary and theoretically possible, using existing measures of (short or long-term) persistence or more sophisticated metrics involving, e.g., function resistance and recovery (Ingrisch and Bahn, 2018). Indeed, in this case, the disturbance is considered to have already taken place in some, if not all, observable systems (Fig. 3). Critically, this means that systems with different resilience to a known disturbance can be compared in terms of baseline biodiversity or statistical properties, and associated resilience indicators either invalidated or confirmed (e.g. Cariveau et al., 2013; Génin et al., 2018; Isbell et al., 2015). Addressing question (i) is essential to increase our ability to predict (anticipate) the consequences of observed environmental change on agroecosystem functions. In contrast, in question (ii), the focus is on resilience to unknown, future disturbances. Observed changes may erode, increase or not affect function resilience to future disturbance (Fig. 4), thus forming a legacy of previous and current changes that influence a function's future responses.

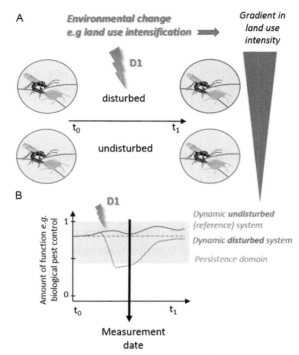

Fig. 3 Breaking down the meaning of an operational framework to measure resilience of agroecosystem functions to environmental change. (A) Biodiversity-driven agroecosystem functions are exemplified by biological pest control taking place in different crop fields of agricultural landscapes. At a hypothetical reference time t_0, crop fields and landscapes are under extensive cultivation (low-intensity cropping practices, small fields, high amounts of non-crop habitat around fields). Between t_0 and t_1, some fields and landscapes undergo varying degrees of intensification (the disturbance D1): fields are enlarged, habitat is cleared, crop management becomes intensive. This leads to a gradient in land use intensity between fields and/or between landscapes. In a space-for-time substitution, pest control in 'undisturbed' landscapes becomes a dynamic reference of the pest control that could have been provided in 'disturbed' landscapes if intensification had not occurred. (B) One aim of an operational resilience framework is to predict pest control resilience to D1 in 'as yet undisturbed' areas, by extrapolating its observed resilience to D1 in disturbed areas. Specifically, we may ask if the function persisted after D1 (stayed in or recovered to a socially acceptable persistence domain), or underwent a regime shift to no or insufficient function. Critically, the temporal resolution of measurements determines their richness of interpretation. Short-term or one-time assessments of persistence, as most often performed, give no indication of potential future recovery. However, because D1 has happened and the function's response can be assessed, these measures are key opportunities to test the performance of indicators of function resilience.

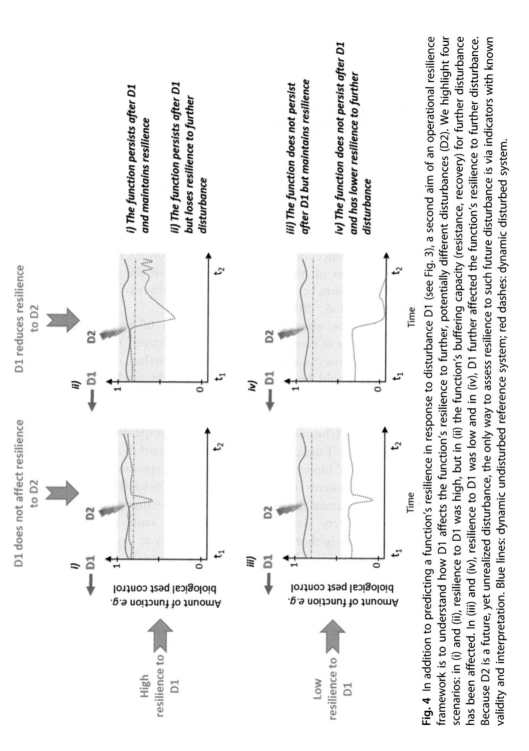

Fig. 4 In addition to predicting a function's resilience in response to disturbance D1 (see Fig. 3), a second aim of an operational resilience framework is to understand how D1 affects the function's resilience to further, potentially different disturbances (D2). We highlight four scenarios: in (i) and (ii), resilience to D1 was high, but in (ii) the function's buffering capacity (resistance, recovery) for further disturbance has been affected. In (iii) and (iv), resilience to D1 was low and in (iv), D1 further affected the function's resilience to further disturbance. Because D2 is a future, yet unrealized disturbance, the only way to assess resilience to such future disturbance is via indicators with known validity and interpretation. Blue lines: dynamic undisturbed reference system; red dashes: dynamic disturbed system.

Since responses to future disturbance are not directly observable, the state of indicators after observed environmental disturbance is a critical tool for their prediction.

5.2 Going beyond: Novel approaches to measuring resilience under environmental change

Current trends and models give indications as to what disturbances can be expected in agroecosystems of different regions, cultivation systems and at different scales (Fig. 2) (Sala et al., 2000). As a result, experimental tests could theoretically go beyond the use of indicators by directly estimating the impact of observed disturbances on resilience to (likely) future disturbance. In practice, this comes down to examining the effects of combinations of known and likely disturbances on the resilience of functioning. Importantly, approaches testing effects of multiple disturbances have the potential to highlight critical interactions between them for the resilience of functioning (e.g. Mantyka-Pringle et al., 2012; Sasaki et al., 2015).

Studies that combine observations of the effects of two or more disturbances on resilience of ecosystem functions are relatively rare, including outside agroecosystems (Sasaki et al., 2015). Nevertheless, some examples reviewed above could be considered in this light, such as studies investigating the impact of local and landscape factors on biological pest control (e.g. Jonsson et al., 2015; Tamburini et al., 2016). In these studies, combinations of a 'press' (chronic) disturbance (e.g. landscape simplification) and a 'pulse' (e.g. tillage; Tamburini et al., 2016) are found to interactively affect the persistence of biological control. Testing experimental 'pulse' disturbances in agroecosystems occurring at the scale of fields (Fig. 2) could be a promising way to attain direct measures of engineering resilience by assessing function recovery rates. Indeed, small-scale, acute disturbances are likely to have relatively short trajectories of recovery at experimentally practicable spatial and temporal scales, and may not require the kind of long-term experiments needed to measure resilience to chronic disturbances occurring at large scales. However, in systems with a legacy of strong large-scale, chronic disturbance (e.g. very simplified landscapes; Tscharntke et al., 2012), we expect the speed of function recovery after small-scale 'pulses' to be slower than in less chronically disturbed systems (Tscharntke et al., 2012). In other words, measuring recovery rates from small-scale disturbances should inform about the relative *engineering* resilience of systems undergoing variable degrees of chronic, large-scale disturbance such as landscape simplification. If we further assume that the validity of slowing recovery rates as indicators for ecological resilience (Table 1; Scheffer et al., 2015) can be extended to biodiversity-driven functions in agroecosystems, then (using a space-for-time substitution; De Palma et al., 2018) these measures may also provide some indication of the functions' *ecological* resilience itself. Although still lacking empirically, this approach has recently been tested theoretically in modelled spatially explicit landscapes, yielding critical insights for practical application in agroecosystems (van de Leemput et al., 2018).

For chronic disturbances such as climate change occurring at a global level, empirical investigations at landscape or regional scales remain difficult. For this, other approaches have recently been applied and may represent promising alternatives. Using a 'trait-for-time' substitution, Kühsel and Blüthgen (2015) examined the resilience of pollinators to climate change by estimating species-specific temperature niches. They show that communities in intensively used landscapes are in fact likely to be more resilient to climate change, as they consist of species (mainly flies) that have broader temperature niches than communities in less disturbed landscapes that have not undergone filtering by land use intensification. While promising, it is still unclear, however, whether such trait-for-time approaches can provide evidence of resilience to climate change of agroecosystem functions (here pollination services) themselves. In this example, resilience of pollination services to further disturbance including climate change may in fact be lower when provided by climate-resilient flies, because it is driven by a community with lower functional diversity (mainly generalists) compared to communities in less disturbed, low-intensity landscapes.

Other approaches to estimate resilience that go beyond some empirical constraints, such as the difficulty to perform experiments and collect data at sufficient spatiotemporal scales, involve statistical and/or theoretical modelling (Egli et al., 2018). For example, modelling of non-linearities in the response of agroecosystem functions to disturbance gradients sometimes uncovers threshold-like changes that may be related to thresholds of ecological resilience (Concepción et al., 2012; With et al., 2002). However, due to the snapshot character of these studies, as discussed above, it is not known whether 'collapsed' functions after such thresholds are on trajectories to recover. In addition, it is also not theoretically known to what extent agroecosystem functions (not communities) are at all subject to thresholds or tipping points of ecological resilience, or instead go through gradual regime shifts when nearing collapse (Dakos et al., 2015). According to theoretical frameworks, communities of function providers outside agroecosystems appear to be subject to tipping points, as shown by recent attempts to model the robustness of communities responsible for providing a function to incremental loss of species. For example, dimension-reduced mutualistic network models have been applied to predict critical tipping points in empirical plant-pollinator communities (Gao et al., 2016; Jiang et al., 2018). In addition, eigenvector decomposition methods are being explored to identify best-indicator species of critical transitions in communities (Dakos, 2018). These examples provide support for the idea that community tipping points also take place in agroecosystems. However, to date, models of community or network resilience have not considered implications in terms of the resilience of associated functions, and appear to generally assume a positive relationship between resilience of a community and resilience of functioning (Jiang et al., 2018). However, if combined with measures of the effectiveness of organisms for service provision in crops, according, for instance, to the degree of trait-matching they exhibit with crops, these methods are promising in estimating resilience of agroecosystem functions to large-scale, chronic environmental disturbances in addition to disturbances at smaller spatial and temporal scales.

5.3 Summary and conclusion: Knowledge gaps and future needs

In this paper, we have reviewed how the resilience of biodiversity-driven functions is, and could be, operationalized in agroecosystems under varying degrees of environmental disturbance. We show that despite a large number of available indicators of resilience, few to date have yet demonstrated validity for biodiversity-driven functions in agroecosystems. Furthermore, studies of three key biodiversity-driven services in agroecosystems highlight that when available, approaches to resilience assessment differ widely between functions, as does the degree to which studies have assessed different dimensions (definitions) of resilience. Overall, we conclude that:

(1) Studies examining the resilience or resilience indicators of biological pest control, disease control in soil, and pollination show variable resilience of these functions to environmental disturbance, but generally low persistence (with exceptions) to land use intensification at local to landscape scales. In contrast, resilience to biotic invasions (pest and pathogen outbreaks, introduction or supplementation of exotic or native pollinators) varies between functions and may depend to a large extent on interactions with land use intensity. Limited data are currently available on the resilience of these functions to climate change, despite progress in assessing the resilience of communities (mainly pollinators) to this driver.

(2) More studies are needed that directly measure or estimate engineering and ecological resilience of agroecosystem functions. Various approaches

developed outside agroecosystems can represent useful blueprints both empirically (direct experimental assessments) and theoretically (indirect estimates and modelling). For this, clarity of concepts is key and we provide a break-down of the steps needed to effectively define the questions being addressed when examining the resilience of biodiversity-driven functions in agroecosystems.

(3) Agroecosystem functions are not independent and disturbances are not occurring in isolation. Assessing the resilience of functions requires consideration of the combined effects of multiple disturbances, including the legacy of previous disturbances. Moreover, trade-offs in the resilience of different functions, or of the same functions at different spatiotemporal scales, are likely to occur and need to be considered if managing agro-ecosystems for resilience of these functions.

(4) Trait-matching between crops and communities of service providers is shown to play a key role in function provision (albeit to date mainly confirmed for pollination), and is likely to strongly influence the resilience of agroecosystem functions. Thus, management for their resilience would benefit from consideration of trait-matching between pools of available service providers and the choice of crops and varieties to be planted. In other words, we need to adapt the crops we cultivate to the means of maintaining resilient functions. Planting crops that benefit from a range of extant biodiversity should enhance the resilience of associated functions. Ultimately, this suggests that maintaining or restoring resilience of agroecosystem functions to environmental change lies to a large extent in our own hands, and effective methods to assess such resilience can provide crucial measures of success.

Acknowledgements

Financial support for this work was provided to M.J. and B.F. through a grant from the Swedish Research Council FORMAS (Grant number 2016-01511, project: 'Will seemingly redundant predator communities maintain stable biological control in the future?'). M.J. and H.F. were financially supported by the Centre for Biological Control at the Swedish University of Agricultural Sciences. E.A.M. and F.R. received no specific grant from any funding agency in the public, commercial, or not-for-profit sectors.

References

Alabouvette, C., Olivain, C., Steinberg, C., 2006a. Biological control of plant diseases: the European situation. Eur. J. Plant Pathol. 114, 329–341. https://doi.org/10.1007/s10658-005-0233-0.

Alabouvette, C., Raaijmakers, J., DE Boer, W., Notz, R., Défago, G., Steinberg, C., Lemanceau, P., 2006b. Concepts and methods to assess the phytosanitary quality of soils. In: Bloem, J., Hopkins, D.W., Benedetti, A. (Eds.), Microbiological Methods for Assessing Soil Quality. CABI Publishing, pp. 257–269.

Allen, C.R., Angeler, D.G., Garmestani, A.S., Gunderson, L.H., Holling, C.S., 2014. Panarchy: theory and application. Ecosystems 17, 578–589. https://doi.org/10.1007/s10021-013-9744-2.

Allen, C.R., Angeler, D.G., Cumming, G.S., Folke, C., Twidwell, D., Uden, D.R., 2016. Quantifying spatial resilience. J. Appl. Ecol. 53, 625–635. https://doi.org/10.1111/1365-2664.12634.

Anderson, S.H., Kelly, D., Ladley, J.J., Molloy, S., Terry, J., 2011. Cascading effects of bird functional extinction reduce pollination and plant density. Science 331, 1068–1071. https://doi.org/10.1126/science.1199092.

Angeler, D.G., Allen, C.R., 2016. Quantifying resilience. J. Appl. Ecol. 53, 617–624. https://doi.org/10.1111/1365-2664.12649.

Asteraki, E.J., Hanks, C.B., Clements, R.O., 1992. The impact of the chemical removal of the hedge-base flora on the community structure of carabid beetles (Col., Carabidae) and spiders (Araneae) of the field and hedge bottom. J. Appl. Entomol. 113, 398–406. https://doi.org/10.1111/j.1439-0418.1992.tb00680.x.

Bailey, R.M., 2010. Spatial and temporal signatures of fragility and threshold proximity in modelled semi-arid vegetation. Proc. R. Soc. Lond. B Biol. Sci. 278, 1064–1071. rspb20101750. https://doi.org/10.1098/rspb.2010.1750.

Baker, K.F., Cook, R.J., 1974. Biological Control of Plant Pathogens. WH Freeman and Company.

Balvanera, P., Pfisterer, A.B., Buchmann, N., He, J.-S., Nakashizuka, T., Raffaelli, D., Schmid, B., 2006. Quantifying the evidence for biodiversity effects on ecosystem functioning and services. Ecol. Lett. 9, 1146–1156. https://doi.org/10.1111/j.1461-0248.2006.00963.x.

Banning, N.C., Murphy, D.V., 2008. Effect of heat-induced disturbance on microbial biomass and activity in forest soil and the relationship between disturbance effects and microbial community structure. Appl. Soil Ecol. 40, 109–119.

Bartomeus, I., Potts, S.G., Steffan-Dewenter, I., Vaissière, B.E., Woyciechowski, M., Krewenka, K.M., Tscheulin, T., Roberts, S.P.M., Szentgyörgyi, H., Westphal, C., Bommarco, R., 2014. Contribution of insect pollinators to crop yield and quality varies with agricultural intensification. PeerJ 2, e328. https://doi.org/10.7717/peerj.328.

Bartomeus, I., Gravel, D., Tylianakis, J.M., Aizen, M.A., Dickie, I.A., Bernard-Verdier, M., 2016. A common framework for identifying linkage rules across different types of interactions. Funct. Ecol. 30, 1894–1903. https://doi.org/10.1111/1365-2435.12666.

Barton, B.T., Ives, A.R., 2014. Species interactions and a chain of indirect effects driven by reduced precipitation. Ecology 95, 486–494. https://doi.org/10.1890/13-0044.1.

Bestelmeyer, B.T., Duniway, M.C., James, D.K., Burkett, L.M., Havstad, K.M., 2013. A test of critical thresholds and their indicators in a desertification-prone ecosystem: more resilience than we thought. Ecol. Lett. 16, 339–345. https://doi.org/10.1111/ele.12045.

Bianchi, F.J., Booij, C.J.H., Tscharntke, T., 2006. Sustainable pest regulation in agricultural landscapes: a review on landscape composition, biodiversity and natural pest control. Proc. R. Soc. Lond. B Biol. Sci. 273, 1715–1727. https://doi.org/10.1098/rspb.2006.3530.

Biesmeijer, J.C., Roberts, S.P.M., Reemer, M., Ohlemüller, R., Edwards, M., Peeters, T., Schaffers, A.P., Potts, S.G., Kleukers, K., Thomas, C.D., Settele, J., Kunin, W.E., 2006. Parallel declines in pollinators and insect-pollinated plants in Britain and the Netherlands. Science 313, 351–354. https://doi.org/10.1126/science.1127863.

Biggs, R., Schlüter, M., Schoon, M.L., 2015. Principles for Building Resilience: Sustaining Ecosystem Services in Social-Ecological Systems. Cambridge University Press.

Birkhofer, K., Bylund, H., Dalin, P., Ferlian, O., Gagic, V., Hambäck, P.A., Klapwijk, M., Mestre, L., Roubinet, E., Schroeder, M., Stenberg, J.A., Porcel, M., Björkman, C., Jonsson, M., 2017. Methods to identify the prey of invertebrate predators in terrestrial field studies. Ecol. Evol. 7, 1942–1953. https://doi.org/10.1002/ece3.2791.

Boetzl, F.A., Krimmer, E., Krauss, J., Steffan-Dewenter, I., 2018. Agri-environmental schemes promote ground-dwelling predators in adjacent oilseed rape fields: diversity, species traits and distance-decay functions. J. Appl. Ecol. 56, 10–20. https://doi.org/10.1111/1365-2664.13162.

Bommarco, R., Marini, L., Vaissière, B.E., 2012. Insect pollination enhances seed yield, quality, and market value in oilseed rape. Oecologia 169, 1025–1032. https://doi.org/10.1007/s00442-012-2271-6.

Bommarco, R., Kleijn, D., Potts, S.G., 2013. Ecological intensification: harnessing ecosystem services for food security. Trends Ecol. Evol. 28, 230–238. https://doi.org/10.1016/j.tree.2012.10.012.

Cabell, J., Oelofse, M., 2012. An indicator framework for assessing agroecosystem resilience. Ecol. Soc. 17, 18. https://doi.org/10.5751/ES-04666-170118.

Cadotte, M.W., Carscadden, K., Mirotchnick, N., 2011. Beyond species: functional diversity and the maintenance of ecological processes and services. J. Appl. Ecol. 48, 1079–1087. https://doi.org/10.1111/j.1365-2664.2011.02048.x.

Cameron, S.A., Lozier, J.D., Strange, J.P., Koch, J.B., Cordes, N., Solter, L.F., Griswold, T.L., 2011. Patterns of widespread decline in North American bumble bees. Proc. Natl. Acad. Sci. U. S. A. 108, 662–667. https://doi.org/10.1073/pnas.1014743108.

Cariveau, D.P., Williams, N.M., Benjamin, F.E., Winfree, R., 2013. Response diversity to land use occurs but does not consistently stabilise ecosystem services provided by native pollinators. Ecol. Lett. 16, 903–911. https://doi.org/10.1111/ele.12126.

Carpenter, S.R., Brock, W.A., 2006. Rising variance: a leading indicator of ecological transition. Ecol. Lett. 9, 311–318. https://doi.org/10.1111/j.1461-0248.2005.00877.x.

Carpenter, S.R., Brock, W.A., 2011. Early warnings of unknown nonlinear shifts: a nonparametric approach. Ecology 92, 2196–2201. https://doi.org/10.1890/11-0716.1.

Carpenter, S., Walker, B., Anderies, J.M., Abel, N., 2001. From metaphor to measurement: resilience of what to what? Ecosystems 4, 765–781. https://doi.org/10.1007/s10021-001-0045-9.

Carvell, C., Meek, W.R., Pywell, R.F., Goulson, D., Nowakowski, M., 2007. Comparing the efficacy of agri-environment schemes to enhance bumble bee abundance and diversity on arable field margins. J. Appl. Ecol. 44, 29–40. https://doi.org/10.1111/j.1365-2664.2006.01249.x.

Case, T.J., 1990. Invasion resistance arises in strongly interacting species-rich model competition communities. Proc. Natl. Acad. Sci. U. S. A. 87, 9610–9614. https://doi.org/10.1073/pnas.87.24.9610.

Caves, E.M., Jennings, S.B., HilleRisLambers, J., Tewksbury, J.J., Rogers, H.S., 2013. Natural experiment demonstrates that bird loss leads to cessation of dispersal of native seeds from intact to degraded forests. PLoS One 8, e65618. https://doi.org/10.1371/journal.pone.0065618.

Chapin, F.S., Walker, B.H., Hobbs, R.J., Hooper, D.U., Lawton, J.H., Sala, O.E., Tilman, D., 1997. Biotic control over the functioning of ecosystems. Science 277, 500–504. https://doi.org/10.1126/science.277.5325.500.

Concepción, E.D., Díaz, M., Kleijn, D., Báldi, A., Batáry, P., Clough, Y., Gabriel, D., Herzog, F., Holzschuh, A., Knop, E., 2012. Interactive effects of landscape context constrain the effectiveness of local agri-environmental management. J. Appl. Ecol. 49, 695–705. https://doi.org/10.1111/j.1365-2664.2012.02131.x.

Cook, R.J., 1993. Making greater use of introduced microorganisms for biological control of plant pathogens. Annu. Rev. Phytopathol. 31, 53–80.

Dai, L., Korolev, K.S., Gore, J., 2013. Slower recovery in space before collapse of connected populations. Nature 496, 355–358. https://doi.org/10.1038/nature12071.

Dakos, V., 2018. Identifying best-indicator species for abrupt transitions in multispecies communities. Ecol. Indic. 94, 494–502. https://doi.org/10.1016/j.ecolind.2017.10.024.

Dakos, V., van Nes, E.H., Donangelo, R., Fort, H., Scheffer, M., 2010. Spatial correlation as leading indicator of catastrophic shifts. Theor. Ecol. 3, 163–174. https://doi.org/ 10.1007/s12080-009-0060-6.

Dakos, V., Kéfi, S., Rietkerk, M., Van Nes, E.H., Scheffer, M., 2011. Slowing down in spatially patterned ecosystems at the brink of collapse. Am. Nat. 177, E153–E166. https://doi.org/10.1086/659945.

Dakos, V., Van Nes, E.H., D'Odorico, P., Scheffer, M., 2012. Robustness of variance and autocorrelation as indicators of critical slowing down. Ecology 93, 264–271. https://doi.org/10.1890/11-0889.1.

Dakos, V., Carpenter, S.R., van Nes, E.H., Scheffer, M., 2015. Resilience indicators: prospects and limitations for early warnings of regime shifts. Philos. Trans. R. Soc. Lond. B Biol. Sci. 370, 20130263. https://doi.org/10.1098/rstb.2013.0263.

Darnhofer, I., Bellon, S., Dedieu, B., Milestad, R., 2010. Adaptiveness to enhance the sustainability of farming systems. A review. Agron. Sustain. Dev. 30, 545–555. https://doi.org/10.1051/agro/2009053.

De Bello, F., Thuiller, W., Lepš, J., Choler, P., Clément, J.-C., Macek, P., Sebastià, M.-T., Lavorel, S., 2009. Partitioning of functional diversity reveals the scale and extent of trait convergence and divergence. J. Veg. Sci. 20, 475–486. https://doi.org/10.1111/j.1654-1103.2009.01042.x.

De Palma, A., Sanchez-Ortiz, K., Martin, P.A., Chadwick, A., Gilbert, G., Bates, A.E., Börger, L., Contu, S., Hill, S.L.L., Purvis, A., 2018. Challenges with inferring how land-use affects terrestrial biodiversity: study design, time, space and synthesis. In: Bohan, D.A., Dumbrell, A.J., Woodward, G., Jackson, M. (Eds.), Next Generation Biomonitoring: Part 1. In: Advances in Ecological Research, vol. 58. Academic Press, pp. 163–199. Chapter 4. https://doi.org/10.1016/bs.aecr.2017.12.004.

Deng, H., 2012. A review of diversity-stability relationship of soil microbial community: what do we not know? J. Environ. Sci. 24, 1027–1035. https://doi.org/10.1016/ S1001-0742(11)60846-2.

Deutsch, C.A., Tewksbury, J.J., Tigchelaar, M., Battisti, D.S., Merrill, S.C., Huey, R.B., Naylor, R.L., 2018. Increase in crop losses to insect pests in a warming climate. Science 361, 916–919. https://doi.org/10.1126/science.aat3466.

Díaz, S., Cabido, M., 2001. Vive la différence: plant functional diversity matters to ecosystem processes. Trends Ecol. Evol. 16, 646–655. https://doi.org/10.1016/S0169-5347(01) 02283-2.

Díaz, S., Purvis, A., Cornelissen, J.H., Mace, G.M., Donoghue, M.J., Ewers, R.M., Jordano, P., Pearse, W.D., 2013. Functional traits, the phylogeny of function, and ecosystem service vulnerability. Ecol. Evol. 3, 2958–2975. https://doi.org/10.1002/ ece3.601.

Diehl, E., Wolters, V., Birkhofer, K., 2012. Arable weeds in organically managed wheat fields foster carabid beetles by resource- and structure-mediated effects. Arthropod Plant Interact. 6, 75–82. https://doi.org/10.1007/s11829-011-9153-4.

Diffenbaugh, N.S., Krupke, C.H., White, M.A., Alexander, C.E., 2008. Global warming presents new challenges for maize pest management. Environ. Res. Lett. 3. 044007 (9pp). https://doi.org/10.1088/1748-9326/3/4/044007.

Donohue, I., Hillebrand, H., Montoya, J.M., Petchey, O.L., Pimm, S.L., Fowler, M.S., Healy, K., Jackson, A.L., Lurgi, M., McClean, D., O'Connor, N.E., O'Gorman, E.J., Yang, Q., 2016. Navigating the complexity of ecological stability. Ecol. Lett. 19, 1172–1185. https://doi.org/10.1111/ele.12648.

Döring, T.F., Vieweger, A., Pautasso, M., Vaarst, M., Finckh, M.R., Wolfe, M.S., 2013. Resilience as a universal criterion of health. J. Sci. Food Agric. 95, 455–465. https:// doi.org/10.1002/jsfa.6539.

Drieu, R., Rusch, A., 2017. Conserving species-rich predator assemblages strengthens natural pest control in a climate warming context. Agric. For. Entomol. 19, 52–59. https://doi.org/10.1111/afe.12180.

Duflot, R., Georges, R., Ernoult, A., Aviron, S., Burel, F., 2014. Landscape heterogeneity as an ecological filter of species traits. Acta Oecol. 56, 19–26. https://doi.org/10.1016/j.actao.2014.01.004.

Dunne, J.A., Williams, R.J., Martinez, N.D., 2002. Network structure and biodiversity loss in food webs: robustness increases with connectance. Ecol. Lett. 5, 558–567. https://doi.org/10.1046/j.1461-0248.2002.00354.x.

Eby, S., Agrawal, A., Majumder, S., Dobson, A.P., Guttal, V., 2017. Alternative stable states and spatial indicators of critical slowing down along a spatial gradient in a savanna ecosystem. Glob. Ecol. Biogeogr. 26, 638–649. https://doi.org/10.1111/geb.12570.

Egli, L., Weise, H., Radchuk, V., Seppelt, R., Grimm, V., 2018. Exploring resilience with agent-based models: state of the art, knowledge gaps and recommendations for coping with multidimensionality. Ecol. Complex. https://doi.org/10.1016/j.ecocom.2018.06.008

Eilenberg, J., Hajek, A., Lomer, C., 2001. Suggestions for unifying the terminology in biological control. BioControl 46, 387–400.

Eisenhauer, N., Schulz, W., Scheu, S., Jousset, A., 2013. Niche dimensionality links biodiversity and invasibility of microbial communities. Funct. Ecol. 27, 282–288. https://doi.org/10.1111/j.1365-2435.2012.02060.x.

Elbgami, T., Kunin, W.E., Hughes, W.O.H., Biesmeijer, J.C., 2014. The effect of proximity to a honeybee apiary on bumblebee colony fitness, development, and performance. Apidologie 45, 504–513. https://doi.org/10.1007/s13592-013-0265-y.

Elead, Y., Köhl, J., Fokkema, N.J., 1994. Control of infection and sporulation of *Botrytis cinerea* on bean and tomato by saprophytic bacteria and fungi. Eur. J. Plant Pathol. 100, 315–336. https://doi.org/10.1007/BF01876443.

Elmqvist, T., Folke, C., Nyström, M., Peterson, G., Bengtsson, J., Walker, B., Norberg, J., 2003. Response diversity, ecosystem change, and resilience. Front. Ecol. Environ. 1, 488–494. https://doi.org/10.1890/1540-9295(2003)001[0488:RDECAR]2.0.CO;2.

Elton, C.S., 1958. The Ecology of Invasions by Animals and Plants. Methuen, London, UK.

European Commission, 2016. Greening. Agriculture and Rural Development. European Commission. URL https://ec.europa.eu/agriculture/direct-support/greening_en (accessed 9.5.18).

Fahrig, L., Baudry, J., Brotons, L., Burel, F.G., Crist, T.O., Fuller, R.J., Sirami, C., Siriwardena, G.M., Martin, J.-L., 2011. Functional landscape heterogeneity and animal biodiversity in agricultural landscapes. Ecol. Lett. 14, 101–112. https://doi.org/10.1111/j.1461-0248.2010.01559.x.

Fargione, J.E., Tilman, D., 2005. Diversity decreases invasion via both sampling and complementarity effects. Ecol. Lett. 8, 604–611. https://doi.org/10.1111/j.1461-0248.2005.00753.x.

Faria, D., Laps, R.R., Baumgarten, J., Cetra, M., 2006. Bat and bird assemblages from forests and shade cacao plantations in two contrasting landscapes in the Atlantic Forest of southern Bahia. Biodivers. Conserv. 15, 587–612. https://doi.org/10.1007/s10531-005-2089-1.

Farwig, N., Schabo, D.G., Albrecht, J., 2017. Trait-associated loss of frugivores in fragmented forest does not affect seed removal rates. J. Ecol. 105, 20–28. https://doi.org/10.1111/1365-2745.12669.

Fitter, A.H., Fitter, R.S.R., 2002. Rapid changes in flowering time in British plants. Science 296, 1689–1691. https://doi.org/10.1126/science.1071617.

Foley, J.A., Ramankutty, N., Brauman, K.A., Cassidy, E.S., Gerber, J.S., Johnston, M., Mueller, N.D., O'Connell, C., Ray, D.K., West, P.C., 2011. Solutions for a cultivated planet. Nature 478, 337. https://doi.org/10.1038/nature10452.

Folke, C., Carpenter, S., Walker, B., Scheffer, M., Elmqvist, T., Gunderson, L., Holling, C.S., 2004. Regime shifts, resilience, and biodiversity in ecosystem management. Annu. Rev. Ecol. Evol. Syst. 35, 557–581. https://doi.org/10.1146/annurev.ecolsys.35.021103.105711.

Folke, C., Carpenter, S.R., Walker, B., Scheffer, M., Chapin, T., Rockström, J., 2010. Resilience thinking: integrating resilience, adaptability and transformability. Ecol. Soc. 15, 20.

Fonseca, C.R., Ganade, G., 2001. Species functional redundancy, random extinctions and the stability of ecosystems. J. Ecol. 89, 118–125. https://doi.org/10.1046/j.1365-2745.2001.00528.x.

Fontaine, C., Dajoz, I., Meriguet, J., Loreau, M., 2005. Functional diversity of plant–pollinator interaction webs enhances the persistence of plant communities. PLoS Biol. 4, e1. https://doi.org/10.1371/journal.pbio.0040001.

Friberg, H., Edel-Hermann, V., Faivre, C., Gautheron, N., Fayolle, L., Faloya, V., Montfort, F., Steinberg, C., 2009. Cause and duration of mustard incorporation effects on soil-borne plant pathogenic fungi. Soil Biol. Biochem. 41, 2075–2084. https://doi.org/10.1016/j.soilbio.2009.07.017.

Frost, C.M., Peralta, G., Rand, T.A., Didham, R.K., Varsani, A., Tylianakis, J.M., 2016. Apparent competition drives community-wide parasitism rates and changes in host abundance across ecosystem boundaries. Nat. Commun. 7, 12644. https://doi.org/10.1038/ncomms12644.

Furlong, M.J., Zalucki, M.P., 2017. Climate change and biological control: the consequences of increasing temperatures on host–parasitoid interactions. Curr. Opin. Insect Sci. 20, 39–44. https://doi.org/10.1016/J.COIS.2017.03.006.

Gabriel, D., Tscharntke, T., 2007. Insect pollinated plants benefit from organic farming. Agric. Ecosyst. Environ. 118, 43–48. https://doi.org/10.1016/j.agee.2006.04.005.

Gagic, V., Hänke, S., Thies, C., Scherber, C., Tomanović, Ž., Tscharntke, T., 2012. Agricultural intensification and cereal aphid-parasitoid-hyperparasitoid food webs: network complexity, temporal variability and parasitism rates. Oecologia 170, 1099–1109. https://doi.org/10.1007/s00442-012-2366-0.

Gagic, V., Bartomeus, I., Jonsson, T., Taylor, A., Winqvist, C., Fischer, C., Slade, E.M., Steffan-Dewenter, I., Emmerson, M., Potts, S.G., Tscharntke, T., Weisser, W., Bommarco, R., 2015. Functional identity and diversity of animals predict ecosystem functioning better than species-based indices. Proc. R. Soc. Lond. B Biol. Sci. 282, 20142620. https://doi.org/10.1098/rspb.2014.2620.

Gagic, V., Kleijn, D., Báldi, A., Boros, G., Jørgensen, H.B., Elek, Z., Garratt, M.P.D., de Groot, G.A., Hedlund, K., Kovács-Hostyánszki, A., Marini, L., Martin, E., Pevere, I., Potts, S.G., Redlich, S., Senapathi, D., Steffan-Dewenter, I., Świtek, S., Smith, H.G., Takács, V., Tryjanowski, P., van der Putten, W.H., van Gils, S., Bommarco, R., 2017. Combined effects of agrochemicals and ecosystem services on crop yield across Europe. Ecol. Lett. 20, 1427–1436. https://doi.org/10.1111/ele.12850.

Gao, J., Barzel, B., Barabási, A.-L., 2016. Universal resilience patterns in complex networks. Nature 530, 307–312. https://doi.org/10.1038/nature16948.

García, D., Martínez, D., Herrera, J.M., Morales, J.M., 2013. Functional heterogeneity in a plant–frugivore assemblage enhances seed dispersal resilience to habitat loss. Ecography 36, 197–208. https://doi.org/10.1111/j.1600-0587.2012.07519.x.

Garibaldi, L.A., Aizen, M.A., Cunningham, S., Klein, A.M., 2009. Pollinator shortage and global crop yield. Commun. Integr. Biol. 2, 37–39. https://doi.org/10.4161/cib.2.1.7425.

Garibaldi, L.A., Bartomeus, I., Bommarco, R., Klein, A.M., Cunningham, S.A., Aizen, M.A., Boreux, V., Garratt, M.P.D., Carvalheiro, L.G., Kremen, C., Morales, C.L., Schüepp, C., Chacoff, N.P., Freitas, B.M., Gagic, V., Holzschuh, A., Klatt, B.K., Krewenka, K.M., Krishnan, S., Mayfield, M.M., Motzke, I., Otieno, M., Petersen, J., Potts, S.G., Ricketts, T.H., Rundlöf, M., Sciligo, A., Sinu, P.A., Steffan-Dewenter, I., Taki, H., Tscharntke, T., Vergara, C.H., Viana, B.F., Woyciechowski, M., Devictor, V., 2015. Editor's choice: review: trait matching of flower visitors and crops predicts fruit set better than trait diversity. J. Appl. Ecol. 52, 1436–1444. https://doi.org/10.1111/1365-2664.12530.

Garibaldi, L.A., Requier, F., Rollin, O., Andersson, G.K.S., 2017. Towards an integrated species and habitat management of crop pollination. Curr. Opin. Insect Sci. 21, 1–10. https://doi.org/10.1016/j.cois.2017.05.016.

Garratt, M.P.D., Breeze, T.D., Jenner, N., Polce, C., Biesmeijer, J.C., Potts, S.G., 2014. Avoiding a bad apple: insect pollination enhances fruit quality and economic value. Agric. Ecosyst. Environ. 184, 34–40. https://doi.org/10.1016/j.agee.2013.10.032.

Geiger, F., Bengtsson, J., Berendse, F., Weisser, W.W., Emmerson, M., Morales, M.B., Ceryngier, P., Liira, J., Tscharntke, T., Winqvist, C., Eggers, S., Bommarco, R., Pärt, T., Bretagnolle, V., Plantegenest, M., Clement, L.W., Dennis, C., Palmer, C., Oñate, J.J., Guerrero, I., Hawro, V., Aavik, T., Thies, C., Flohre, A., Hänke, S., Fischer, C., Goedhart, P.W., Inchausti, P., 2010. Persistent negative effects of pesticides on biodiversity and biological control potential on European farmland. Basic Appl. Ecol. 11, 97–105. https://doi.org/10.1016/j.baae.2009.12.001.

Geldmann, J., González-Varo, J.P., 2018. Conserving honey bees does not help wildlife. Science 359, 392–393. https://doi.org/10.1126/science.aar2269.

Génin, A., Majumder, S., Sankaran, S., Schneider, F.D., Danet, A., Berdugo, M., Guttal, V., Kéfi, S., 2018. Spatially heterogeneous stressors can alter the performance of indicators of regime shifts. Ecol. Indic. 94, 520–533. https://doi.org/10.1016/j.ecolind.2017.10.071.

Geslin, B., Gauzens, B., Baude, M., Dajoz, I., Fontaine, C., Henry, M., Ropars, L., Rollin, O., Thébault, E., Vereecken, N.J., 2017. Massively introduced managed species and their consequences for plant–pollinator interactions. In: Networks of Invasion: Empirical Evidence and Case Studies. Advances in Ecological Research, vol. 57. Academic Press, pp. 147–199. Chapter 4. https://doi.org/10.1016/bs.aecr.2016.10.007.

González-Varo, J.P., Carvalho, C.S., Arroyo, J.M., Jordano, P., 2017. Unravelling seed dispersal through fragmented landscapes: frugivore species operate unevenly as mobile links. Mol. Ecol. 26, 4309–4321. https://doi.org/10.1111/mec.14181.

Gossner, M.M., Simons, N.K., Achtziger, R., Blick, T., Dorow, W.H., Dziock, F., Köhler, F., Rabitsch, W., Weisser, W.W., 2015. A summary of eight traits of Coleoptera, Hemiptera, Orthoptera and Araneae, occurring in grasslands in Germany. Sci. Data 2, 150013.

Goulson, D., Sparrow, K.R., 2009. Evidence for competition between honeybees and bumblebees; effects on bumblebee worker size. J. Insect Conserv. 13, 177–181.

Goulson, D., Lye, G.C., Darvill, B., 2008. Decline and conservation of bumble bees. Annu. Rev. Entomol. 53, 191–208.

Goulson, D., Nicholls, E., Botias, C., Rotheray, E.L., 2015. Bee declines driven by combined stress from parasites, pesticides, and lack of flowers. Science 347, 1255957.

Griffiths, B.S., Philippot, L., 2013. Insights into the resistance and resilience of the soil microbial community. FEMS Microbiol. Rev. 37, 112–129. https://doi.org/10.1111/j.1574-6976.2012.00343.x.

Griffiths, B., Ritz, K., Wheatley, R., Kuan, H., Boag, B., Christensen, S., Ekelund, F., Sørensen, S., Muller, S., Bloem, J., 2001. An examination of the biodiversity–ecosystem function relationship in arable soil microbial communities. Soil Biol. Biochem. 33, 1713–1722. https://doi.org/10.1016/S0038-0717(01)00094-3.

Grime, J.P., 1998. Benefits of plant diversity to ecosystems: immediate, filter and founder effects. J. Ecol. 86, 902–910. https://doi.org/10.1046/j.1365-2745.1998.00306.x.

Grimm, V., Wissel, C., 1997. Babel, or the ecological stability discussions: an inventory and analysis of terminology and a guide for avoiding confusion. Oecologia 109, 323–334. https://doi.org/10.1007/s004420050090.

Guetsky, R., Shtienberg, D., Elad, Y., Dinoor, A., 2001. Combining biocontrol agents to reduce the variability of biological control. Phytopathology 91, 621–627. https://doi.org/10.1094/PHYTO.2001.91.7.621.

Hallett, A.C., Mitchell, R.J., Chamberlain, E.R., Karron, J.D., 2017. Pollination success following loss of a frequent pollinator: the role of compensatory visitation by other effective pollinators. AoB Plants 9, plx020.

Hasna, M.K., Mårtensson, A., Persson, P., Rämert, B., 2007. Use of composts to manage corky root disease in organic tomato production. Ann. Appl. Biol. 151, 381–390. https://doi.org/10.1111/j.1744-7348.2007.00178.x.

Henry, M., Beguin, M., Requier, F., Rollin, O., Odoux, J.F., Aupinel, P., Aptel, J., Tchamitchian, S., Decourtye, A., 2012. A common pesticide decreases foraging success and survival in honey bees. Science 336, 348–350.

Hillebrand, H., 2004. On the generality of the latitudinal diversity gradient. Am. Nat. 163, 192–211. https://doi.org/10.1086/381004.

Ho, W.C., Ko, W.H., 1982. Characteristics of soil microbiostasis. Soil Biol. Biochem. 14, 589–593. https://doi.org/10.1016/0038-0717(82)90092-X.

Hodgson, D., McDonald, J.L., Hosken, D.J., 2015. What do you mean, 'resilient'? Trends Ecol. Evol. 30, 503–506. https://doi.org/10.1016/j.tree.2015.06.010.

Holling, C.S., 1973. Resilience and stability of ecological systems. Annu. Rev. Ecol. Syst. 4, 1–23.

Holling, C.S., 1988. Temperate forest insect outbreaks, tropical deforestation and migratory birds. Mem. Entomol. Soc. Can. 120, 21–32. https://doi.org/10.4039/entm120146021-1.

Holzschuh, A., Steffan-Dewenter, I., Kleijn, D., Tscharntke, T., 2007. Diversity of flower-visiting bees in cereal fields: effects of farming system, landscape composition and regional context. J. Appl. Ecol. 44, 41–49.

Holzschuh, A., Dudenhöffer, J.H., Tscharntke, T., 2012. Landscapes with wild bee habitats enhance pollination, fruit set and yield of sweet cherry. Biol. Conserv. 153, 101–107.

Hooper, D.U., Chapin, F.S., Ewel, J.J., Hector, A., Inchausti, P., Lavorel, S., Lawton, J.H., Lodge, D.M., Loreau, M., Naeem, S., Schmid, B., Setala, H., Symstad, A.J., Vandermeer, J., Wardle, D.A., 2005. Effects of biodiversity on ecosystem functioning: a consensus of current knowledge. Ecol. Monogr. 75, 3–35. https://doi.org/10.1890/04-0922.

Hooper, D.U., Adair, E.C., Cardinale, B.J., Byrnes, J.E.K., Hungate, B.A., Matulich, K.L., Gonzalez, A., Duffy, J.E., Gamfeldt, L., O'Connor, M.I., 2012. A global synthesis reveals biodiversity loss as a major driver of ecosystem change. Nature 486, 105–108. https://doi.org/10.1038/nature11118.

Houlahan, J.E., Currie, D.J., Cottenie, K., Cumming, G.S., Ernest, S.K.M., Findlay, C.S., Fuhlendorf, S.D., Gaedke, U., Legendre, P., Magnuson, J.J., McArdle, B.H., Muldavin, E.H., Noble, D., Russell, R., Stevens, R.D., Willis, T.J., Woiwod, I.P., Wondzell, S.M., 2007. Compensatory dynamics are rare in natural ecological communities. Proc. Natl. Acad. Sci. U. S. A. 104, 3273–3277. https://doi.org/10.1073/pnas.0603798104.

Hudewenz, A., Klein, A.M., 2013. Competition between honey bees and wild bees and the role of nesting resources in a nature reserve. J. Insect Conserv. 17, 1275–1283.

Ingrisch, J., Bahn, M., 2018. Towards a comparable quantification of resilience. Trends Ecol. Evol. 33, 251–259. https://doi.org/10.1016/j.tree.2018.01.013.

Irikiin, Y., Nishiyama, M., Otsuka, S., Senoo, K., 2006. Rhizobacterial community-level, sole carbon source utilization pattern affects the delay in the bacterial wilt of tomato grown in rhizobacterial community model system. Appl. Soil Ecol. 34, 27–32. https://doi.org/10.1016/j.apsoil.2005.12.003.

Isaacs, R., Williams, N., Ellis, J., Pitts-Singer, T.L., Bommarco, R., Vaughan, M., 2017. Integrated crop pollination: combining strategies to ensure stable and sustainable yields of pollination-dependent crops. Basic Appl. Ecol. 22, 44–60.

Isbell, F., Craven, D., Connolly, J., Loreau, M., Schmid, B., Beierkuhnlein, C., Bezemer, T.M., Bonin, C., Bruelheide, H., de Luca, E., Ebeling, A., Griffin, J.N., Guo, Q., Hautier, Y., Hector, A., Jentsch, A., Kreyling, J., Lanta, V., Manning, P., Meyer, S.T., Mori, A.S., Naeem, S., Niklaus, P.A., Polley, H.W., Reich, P.B., Roscher, C., Seabloom, E.W., Smith, M.D., Thakur, M.P., Tilman, D., Tracy, B.F., van der Putten, W.H., van Ruijven, J., Weigelt, A., Weisser, W.W., Wilsey, B., Eisenhauer, N., 2015. Biodiversity increases the resistance of ecosystem productivity to climate extremes. Nature 526, 574–577. https://doi.org/10.1038/nature15374.

Jiang, J., Huang, Z.-G., Seager, T.P., Lin, W., Grebogi, C., Hastings, A., Lai, Y.-C., 2018. Predicting tipping points in mutualistic networks through dimension reduction. Proc. Natl. Acad. Sci. U. S. A. 115, E639–E647. https://doi.org/10.1073/pnas.1714958115.

Joner, F., Specht, G., Müller, S.C., Pillar, V.D., 2011. Functional redundancy in a clipping experiment on grassland plant communities. Oikos 120, 1420–1426. https://doi.org/10.1111/j.1600-0706.2011.19375.x.

Jonsson, M., Straub, C.S., Didham, R.K., Buckley, H.L., Case, B.S., Hale, R.J., Gratton, C., Wratten, S.D., 2015. Experimental evidence that the effectiveness of conservation biological control depends on landscape complexity. J. Appl. Ecol. 52, 1274–1282. https://doi.org/10.1111/1365-2664.12489.

Jonsson, M., Kaartinen, R., Straub, C.S., 2017. Relationships between natural enemy diversity and biological control. Curr. Opin. Insect Sci. 20, 1–6. https://doi.org/10.1016/j.cois.2017.01.001.

Kalda, O., Kalda, R., Liira, J., 2015. Multi-scale ecology of insectivorous bats in agricultural landscapes. Agric. Ecosyst. Environ. 199, 105–113. https://doi.org/10.1016/j.agee.2014.08.028.

Karp, D.S., Ziv, G., Zook, J., Ehrlich, P.R., Daily, G.C., 2011. Resilience and stability in bird guilds across tropical countryside. Proc. Natl. Acad. Sci. U. S. A. 108, 21134–21139. https://doi.org/10.1073/pnas.1118276108.

Karp, D.S., Chaplin-Kramer, R., Meehan, T.D., Martin, E.A., DeClerck, F., Grab, H., Gratton, C., Hunt, L., Larsen, A.E., Martínez-Salinas, A., O'Rourke, M.E., Rusch, A., Poveda, K., Jonsson, M., Rosenheim, J.A., Schellhorn, N.A., Tscharntke, T., Wratten, S.D., Zhang, W., Iverson, A.L., Adler, L.S., Albrecht, M., Alignier, A., Angelella, G.M., Zubair Anjum, M., Avelino, J., Batáry, P., Baveco, J.M., Bianchi, F.J.J.A., Birkhofer, K., Bohnenblust, E.W., Bommarco, R., Brewer, M.J., Caballero-López, B., Carrière, Y., Carvalheiro, L.G., Cayuela, L., Centrella, M., Ćetković, A., Henri, D.C., Chabert, A., Costamagna, A.C., De la Mora, A., de Kraker, J., Desneux, N., Diehl, E., Diekötter, T., Dormann, C.F., Eckberg, J.O., Entling, M.H., Fiedler, D., Franck, P., Frank van Veen, F.J., Frank, T., Gagic, V., Garratt, M.P.D., Getachew, A., Gonthier, D.J., Goodell, P.B., Graziosi, I., Groves, R.L., Gurr, G.M., Hajian-Forooshani, Z., Heimpel, G.E., Herrmann, J.D., Huseth, A.S., Inclán, D.J., Ingrao, A.J., Iv, P., Jacot, K., Johnson, G.A., Jones, L., Kaiser, M., Kaser, J.M., Keasar, T., Kim, T.N., Kishinevsky, M., Landis, D.A., Lavandero, B., Lavigne, C., Le Ralec, A., Lemessa, D., Letourneau, D.K., Liere, H., Lu, Y., Lubin, Y., Luttermoser, T., Maas, B., Mace, K., Madeira, F., Mader, V., Cortesero, A.M., Marini, L., Martinez, E., Martinson, H.M., Menozzi, P., Mitchell, M.G.E., Miyashita, T., Molina, G.A.R., Molina-Montenegro, M.A.,

O'Neal, M.E., Opatovsky, I., Ortiz-Martinez, S., Nash, M., Östman, Ö., Ouin, A., Pak, D., Paredes, D., Parsa, S., Parry, H., Perez-Alvarez, R., Perović, D.J., Peterson, J.A., Petit, S., Philpott, S.M., Plantegenest, M., Plećaš, M., Pluess, T., Pons, X., Potts, S.G., Pywell, R.F., Ragsdale, D.W., Rand, T.A., Raymond, L., Ricci, B., Sargent, C., Sarthou, J.-P., Saulais, J., Schäckermann, J., Schmidt, N.P., Schneider, G., Schüepp, C., Sivakoff, F.S., Smith, H.G., Stack Whitney, K., Stutz, S., Szendrei, Z., Takada, M.B., Taki, H., Tamburini, G., Thomson, L.J., Tricault, Y., Tsafack, N., Tschumi, M., Valantin-Morison, M., Van Trinh, M., van der Werf, W., Vierling, K.T., Werling, B.P., Wickens, J.B., Wickens, V.J., Woodcock, B.A., Wyckhuys, K., Xiao, H., Yasuda, M., Yoshioka, A., Zou, Y., 2018. Crop pests and predators exhibit inconsistent responses to surrounding landscape composition. Proc. Natl. Acad. Sci. U. S. A. 115, E7863–E7870. https://doi.org/10.1073/pnas.1800042115.

Kennedy, C.M., Lonsdorf, E., Neel, M.C., Williams, N.M., Ricketts, T.H., Winfree, R., Bommarco, R., Brittain, C., Burley, A.L., Cariveau, D., Carvalheiro, L.G., Chacoff, N.P., Cunningham, S.A., Danforth, B.N., Dudenhöffer, J.H., Elle, E., Gaines, H.R., Garibaldi, L.A., Gratton, C., Holzschuh, A., Isaacs, R., Javorek, S.K., Jha, S., Klein, A.M., Krewenka, K., Mandelik, Y., Mayfield, M.M., Morandin, L., Neame, L.A., Otieno, M., Park, M., Potts, S.G., Rundlöf, M., Saez, A., Steffan-Dewenter, I., Taki, H., Viana, B.F., Westphal, C., Wilson, J.K., Greenleaf, S.S., Kremen, C., 2013. A global quantitative synthesis of local and landscape effects on wild bee pollinators in agroecosystems. Ecol. Lett. 16, 584–599.

Kinkel, L.L., Bakker, M.G., Schlatter, D.C., 2011. A coevolutionary framework for managing disease-suppressive soils. Annu. Rev. Phytopathol. 49, 47–67. https://doi.org/10.1146/annurev-phyto-072910-095232.

Klein, A.-M., Vaissiere, B.E., Cane, J.H., Steffan-Dewenter, I., Cunningham, S.A., Kremen, C., Tscharntke, T., 2007. Importance of pollinators in changing landscapes for world crops. Proc. R. Soc. Lond. B Biol. Sci. 274, 303–313.

Korenko, S., Niedobová, J., Kolářová, M., Hamouzová, K., Kysilková, K., Michalko, R., 2016. The effect of eight common herbicides on the predatory activity of the agrobiont spider Pardosa agrestis. BioControl 61, 507–517. https://doi.org/10.1007/s10526-016-9729-0.

Krauss, J., Gallenberger, I., Steffan-Dewenter, I., 2011. Decreased functional diversity and biological pest control in conventional compared to organic crop fields. PLoS One 6, e19502. https://doi.org/10.1371/journal.pone.0019502.

Kremen, C., M'Gonigle, L.K., 2015. Small-scale restoration in intensive agricultural landscapes supports more specialized and less mobile pollinator species. J. Appl. Ecol. 52, 602–610.

Kremen, C., Williams, N.M., Thorp, R.W., 2002. Crop pollination from native bees at risk from agricultural intensification. Proc. Natl. Acad. Sci. U. S. A. 99, 16812–16816.

Kruess, A., Tscharntke, T., 1994. Habitat fragmentation, species loss, and biological control. Science 264, 1581–1584. https://doi.org/10.1126/science.264.5165.1581.

Kruess, A., Tscharntke, T., 2000. Species richness and parasitism in a fragmented landscape: experiments and field studies with insects on Vicia sepium. Oecologia 122, 129–137. https://doi.org/10.1007/PL00008829.

Kühsel, S., Blüthgen, N., 2015. High diversity stabilizes the thermal resilience of pollinator communities in intensively managed grasslands. Nat. Commun. 6, 7989. https://doi.org/10.1038/ncomms8989.

Laliberté, E., Tylianakis, J.M., 2010. Deforestation homogenizes tropical parasitoid-host networks. Ecology 91, 1740–1747. https://doi.org/10.1890/09-1328.1.

Landis, D.A., Wratten, S.D., Gurr, G.M., 2000. Habitat management to conserve natural enemies of arthropod pests in agriculture. Annu. Rev. Entomol. 45, 175–201. https://doi.org/10.1146/annurev.ento.45.1.175.

Larkin, R., 1996. Suppression of Fusarium wilt of watermelon by nonpathogenic *Fusarium oxysporum* and other microorganisms recovered from a disease-suppressive soil. Phytopathology 86, 812. https://doi.org/10.1094/Phyto-86-812.

Larkin, R.P., 2015. Soil health paradigms and implications for disease management. Annu. Rev. Phytopathol. 53, 199–221. https://doi.org/10.1146/annurev-phyto-080614-120357.

Larsen, T.H., Williams, N.M., Kremen, C., 2005. Extinction order and altered community structure rapidly disrupt ecosystem functioning. Ecol. Lett. 8, 538–547. https://doi.org/10.1111/j.1461-0248.2005.00749.x.

Latz, E., Eisenhauer, N., Rall, B.C., Scheu, S., Jousset, A., 2016. Unravelling linkages between plant community composition and the pathogen-suppressive potential of soils. Sci. Rep. 6, 23584. https://doi.org/10.1038/srep23584.

Lavorel, S., Garnier, E., 2002. Predicting changes in community composition and ecosystem functioning from plant traits: revisiting the Holy Grail. Funct. Ecol. 16, 545–556. https://doi.org/10.1046/j.1365-2435.2002.00664.x.

Le Féon, V., Schermann-Legionnet, A., Delettre, Y., Aviron, S., Billeter, R., Bugter, R., Hendrickx, F., Burel, F., 2010. Intensification of agriculture, landscape composition and wild bee communities: a large scale study in four European countries. Agric. Ecosyst. Environ. 137, 143–150.

Le Féon, V., Burel, F., Chifflet, R., Henry, M., Ricroch, A., Vaissière, B.E., Baudry, J., 2013. Solitary bee abundance and species richness in dynamic agricultural landscapes. Agric. Ecosyst. Environ. 166, 94–101.

Lechenet, M., Dessaint, F., Py, G., Makowski, D., Munier-Jolain, N., 2017. Reducing pesticide use while preserving crop productivity and profitability on arable farms. Nat. Plants 3, 17008. https://doi.org/10.1038/nplants.2017.8.

Lever, J.J., van Nes, E.H., Scheffer, M., Bascompte, J., 2014. The sudden collapse of pollinator communities. Ecol. Lett. 17, 350–359. https://doi.org/10.1111/ele.12236.

Levine, J.M., Adler, P.B., Yelenik, S.G., 2004. A meta-analysis of biotic resistance to exotic plant invasions. Ecol. Lett. 7, 975–989. https://doi.org/10.1111/j.1461-0248.2004.00657.x.

Lockwood, J.L., 1977. Fungistasis in soils. Biol. Rev. 52, 1–43. https://doi.org/10.1111/j.1469-185X.1977.tb01344.x.

Lövei, G.L., Ferrante, M., 2017. A review of the sentinel prey method as a way of quantifying invertebrate predation under field conditions. Insect Sci. 24, 528–542.

Lundberg, J., Moberg, F., 2003. Mobile link organisms and ecosystem functioning: implications for ecosystem resilience and management. Ecosystems 6, 0087–0098. https://doi.org/10.1007/s10021-002-0150-4.

Maas, B., Tscharntke, T., Saleh, S., Dwi Putra, D., Clough, Y., 2015. Avian species identity drives predation success in tropical cacao agroforestry. J. Appl. Ecol. 52, 735–743. https://doi.org/10.1111/1365-2664.12409.

Macfadyen, S., Gibson, R., Polaszek, A., Morris, R.J., Craze, P.G., Planqué, R., Symondson, W.O.C., Memmott, J., 2009. Do differences in food web structure between organic and conventional farms affect the ecosystem service of pest control? Ecol. Lett. 12, 229–238. https://doi.org/10.1111/j.1461-0248.2008.01279.x.

Macfadyen, S., Craze, P.G., Polaszek, A., van Achterberg, K., Memmott, J., 2011. Parasitoid diversity reduces the variability in pest control services across time on farms. Proc. R. Soc. B Biol. Sci. 278, 3387–3394. https://doi.org/10.1098/rspb.2010.2673.

Maiorano, A., Cerrani, I., Fumagalli, D., Donatelli, M., 2014. New biological model to manage the impact of climate warming on maize corn borers. Agron. Sustain. Dev. 34, 609–621. https://doi.org/10.1007/s13593-013-0185-2.

Mantyka-Pringle, C.S., Martin, T.G., Rhodes, J.R., 2012. Interactions between climate and habitat loss effects on biodiversity: a systematic review and meta-analysis. Glob. Chang. Biol. 18, 1239–1252.

Martin, E.A., Reineking, B., Seo, B., Steffan-Dewenter, I., 2013. Natural enemy interactions constrain pest control in complex agricultural landscapes. Proc. Natl. Acad. Sci. 110, 5534–5539. https://doi.org/10.1073/pnas.1215725110.

Martin, E.A., Seo, B., Park, C.-R., Reineking, B., Steffan-Dewenter, I., 2016. Scale-dependent effects of landscape composition and configuration on natural enemy diversity, crop herbivory, and yields. Ecol. Appl. 26, 448–462. https://doi.org/10.1890/15-0856.1.

Matos, A., Kerkhof, L., Garland, J.L., 2005. Effects of microbial community diversity on the survival of Pseudomonas aeruginosa in the wheat rhizosphere. Microb. Ecol. 49, 257–264. https://doi.org/10.1007/s00248-004-0179-3.

Matthiessen, J.N., Kirkegaard, J.A., 2006. Biofumigation and enhanced biodegradation: opportunity and challenge in soilborne pest and disease management. Crit. Rev. Plant Sci. 25, 235–265. https://doi.org/10.1080/07352680600611543.

Mazzola, M., Brown, J., Izzo, A.D., Cohen, M.F., 2007. Mechanism of action and efficacy of seed meal-induced pathogen suppression differ in a Brassicaceae species and time-dependent manner. Phytopathology 97, 454–460. https://doi.org/10.1094/PHYTO-97-4-0454.

McGill, B.J., Enquist, B.J., Weiher, E., Westoby, M., 2006. Rebuilding community ecology from functional traits. Trends Ecol. Evol. 21, 178–185. https://doi.org/10.1016/j.tree.2006.02.002.

McKey, D., Rostain, S., Iriarte, J., Glaser, B., Birk, J.J., Holst, I., Renard, D., 2010. Pre-Columbian agricultural landscapes, ecosystem engineers, and self-organized patchiness in Amazonia. Proc. Natl. Acad. Sci. U. S. A. 107, 7823–7828. https://doi.org/10.1073/pnas.0908925107.

Menalled, F.D., Costamagna, A.C., Marino, P.C., Landis, D.A., 2003. Temporal variation in the response of parasitoids to agricultural landscape structure. Agric. Ecosyst. Environ. 96, 29–35.

Millennium Ecosystem Assessment, 2005. Ecosystems and Human Well-Being: Current State and Trends. Island Press, Washington, D.C.

Mocali, S., Landi, S., Curto, G., Dallavalle, E., Infantino, A., Colzi, C., d'Errico, G., Roversi, P.F., D'Avino, L., Lazzeri, L., 2015. Resilience of soil microbial and nematode communities after biofumigant treatment with defatted seed meals. Ind. Crop. Prod. 75, 79–90. https://doi.org/10.1016/j.indcrop.2015.04.031.

Mommer, L., Anne Cotton, T.E., Raaijmakers, J.M., Termorshuizen, A.J., van Ruijven, J., Hendriks, M., van Rijssel, S.Q., van de Mortel, J.E., van der Paauw, J.W., Schijlen, E.G.W.M., Smit-Tiekstra, A.E., Berendse, F., de Kroon, H., Dumbrell, A.J., 2018. Lost in diversity: the interactions between soil-borne fungi, biodiversity and plant productivity. New Phytol. 218, 542–553. https://doi.org/10.1111/nph.15036.

Mora, F., 2017. A structural equation modeling approach for formalizing and evaluating ecological integrity in terrestrial ecosystems. Eco. Inform. 41, 74–90. https://doi.org/10.1016/j.ecoinf.2017.05.002.

Morales, C.L., Arbetman, M.P., Cameron, S.A., Aizen, M.A., 2013. Rapid ecological replacement of a native bumble bee by invasive species. Front. Ecol. Environ. 11, 529–534.

Moreira, J.I., Riba-Hernández, P., Lobo, J.A., 2017. Toucans (Ramphastos ambiguus) facilitate resilience against seed dispersal limitation to a large-seeded tree (Virola surinamensis) in a human-modified landscape. Biotropica 49, 502–510. https://doi.org/10.1111/btp.12427.

Mori, A.S., 2016. Resilience in the studies of biodiversity–ecosystem functioning. Trends Ecol. Evol. 31, 87–89. https://doi.org/10.1016/j.tree.2015.12.010.

Mori, A.S., Furukawa, T., Sasaki, T., 2013. Response diversity determines the resilience of ecosystems to environmental change. Biol. Rev. 88, 349–364. https://doi.org/10.1111/brv.12004.

Naeem, S., 1998. Species redundancy and ecosystem reliability. Conserv. Biol. 12, 39–45. https://doi.org/10.1111/j.1523-1739.1998.96379.x.

Naeem, S., Wright, J.P., 2003. Disentangling biodiversity effects on ecosystem functioning: deriving solutions to a seemingly insurmountable problem. Ecol. Lett. 6, 567–579. https://doi.org/10.1046/j.1461-0248.2003.00471.x.

Nash, K.L., Graham, N.A., Jennings, S., Wilson, S.K., Bellwood, D.R., 2016. Herbivore cross-scale redundancy supports response diversity and promotes coral reef resilience. J. Appl. Ecol. 53, 646–655.

Norberg, J., Swaney, D.P., Dushoff, J., Lin, J., Casagrandi, R., Levin, S.A., 2001. Phenotypic diversity and ecosystem functioning in changing environments: a theoretical framework. Proc. Natl. Acad. Sci. U. S. A. 98, 11376–11381. https://doi.org/10.1073/pnas.171315998.

Nurdiansyah, F., Denmead, L.H., Clough, Y., Wiegand, K., Tscharntke, T., 2016. Biological control in Indonesian oil palm potentially enhanced by landscape context. Agric. Ecosyst. Environ. 232, 141–149. https://doi.org/10.1016/j.agee.2016.08.006.

Nyffeler, M., Sterling, W.L., Dean, D.A., 1994. How spiders make a living. Environ. Entomol. 23, 1357–1367. https://doi.org/10.1093/ee/23.6.1357.

Öckinger, E., Schweiger, O., Crist, T.O., Debinski, D.M., Krauss, J., Kuussaari, M., Petersen, J.D., Pöyry, J., Settele, J., Summerville, K.S., Bommarco, R., 2010. Life-history traits predict species responses to habitat area and isolation: a cross-continental synthesis. Ecol. Lett. 13, 969–979. https://doi.org/10.1111/j.1461-0248.2010.01487.x.

Oerke, E.-C., 2006. Crop losses to pests. J. Agric. Sci. 144, 31–43.

Okuyama, T., Holland, J.N., 2008. Network structural properties mediate the stability of mutualistic communities. Ecol. Lett. 11, 208–216. https://doi.org/10.1111/j.1461-0248.2007.01137.x.

Oliver, T.H., Heard, M.S., Isaac, N.J.B., Roy, D.B., Procter, D., Eigenbrod, F., Freckleton, R., Hector, A., Orme, C.D.L., Petchey, O.L., Proença, V., Raffaelli, D., Suttle, K.B., Mace, G.M., Martín-López, B., Woodcock, B.A., Bullock, J.M., 2015a. Biodiversity and resilience of ecosystem functions. Trends Ecol. Evol. 30, 673–684. https://doi.org/10.1016/j.tree.2015.08.009.

Oliver, T.H., Isaac, N.J.B., August, T.A., Woodcock, B.A., Roy, D.B., Bullock, J.M., 2015b. Declining resilience of ecosystem functions under biodiversity loss. Nat. Commun. 6, 10122. https://doi.org/10.1038/ncomms10122.

Ollerton, J., Winfree, R., Tarrant, S., 2011. How many flowering plants are pollinated by animals? Oikos 120, 321–326.

Orwin, K.H., Wardle, D.A., Greenfield, L.G., 2006. Context-dependent changes in the resistance and resilience of soil microbes to an experimental disturbance for three primary plant chronosequences. Oikos 112, 196–208.

Östman, Ö., Ekbom, B., Bengtsson, J., 2003. Yield increase attributable to aphid predation by ground-living polyphagous natural enemies in spring barley in Sweden. Ecol. Econ. 45, 149–158. https://doi.org/10.1016/S0921-8009(03)00007-7.

Partap, U., Ya, T., 2012. The human pollinators of fruit crops in Maoxian county, Sichuan, China. Mt. Res. Dev. 32, 176–186.

Peralta, G., Frost, C.M., Rand, T.A., Didham, R.K., Tylianakis, J.M., 2014. Complementarity and redundancy of interactions enhance attack rates and spatial stability in host–parasitoid food webs. Ecology 95, 1888–1896. https://doi.org/10.1890/13-1569.1.

Pérez-Piqueres, A., Edel-Hermann, V., Alabouvette, C., Steinberg, C., 2006. Response of soil microbial communities to compost amendments. Soil Biol. Biochem. 38, 460–470. https://doi.org/10.1016/j.soilbio.2005.05.025.

Perfecto, I., Vandermeer, J.H., Bautista, G.L., Nuñez, G.I., Greenberg, R., Bichier, P., Langridge, S., 2004. Greater predation in shaded coffee farms: the role of resident neotropical birds. Ecology 85, 2677–2681. https://doi.org/10.1890/03-3145.

Peterson, G., Allen, C.R., Holling, C.S., 1998. Ecological resilience, biodiversity, and scale. Ecosystems 1, 6–18.

Peterson, C.A., Eviner, V.T., Gaudin, A.C., 2018. Ways forward for resilience research in agroecosystems. Agr. Syst. 162, 19–27.

Pillar, V.D., Blanco, C.C., Müller, S.C., Sosinski, E.E., Joner, F., Duarte, L.D., 2013. Functional redundancy and stability in plant communities. J. Veg. Sci. 24, 963–974.

Pisa, L.W., Amaral-Rogers, V., Belzunces, L.P., Bonmatin, J.M., Downs, C.A., Goulson, D., Kreutzweiser, D.P., Krupke, C., Liess, M., McField, M., Morrissey, C.A., Noome, D.A., Settele, J., Simon-Delso, N., Stark, J.D., Van der Sluijs, J.P., Van Dyck, H., Wiemers, M., 2015. Effects of neonicotinoids and fipronil on nontarget invertebrates. Environ. Sci. Pollut. Res. 22, 68–102. https://doi.org/10.1007/s11356-014-3471-x.

Pitts-Singer, T.L., Artz, D.R., Peterson, S.S., Boyle, N.K., Wardell, G.I., 2018. Examination of a managed pollinator strategy for almond production using *Apis mellifera* (Hymenoptera: Apidae) and *Osmia lignaria* (Hymenoptera: Megachilidae). Environ. Entomol. 47, 364–377. https://doi.org/10.1093/ee/nvy009.

Poisot, T., Mouquet, N., Gravel, D., 2013. Trophic complementarity drives the biodiversity-ecosystem functioning relationship in food webs. Ecol. Lett. 16, 853–861. https://doi.org/10.1111/ele.12118.

Porcel, M., Andersson, G.K.S., Pålsson, J., Tasin, M., 2018. Organic management in apple orchards: higher impacts on biological control than on pollination. J. Appl. Ecol. 55, 2779–2789. https://doi.org/10.1111/1365-2664.13247.

Potts, S.G., Biesmeijer, J.C., Kremen, C., Neumann, P., Schweiger, O., Kunin, W.E., 2010. Global pollinator declines: trends, impacts and drivers. Trends Ecol. Evol. 25, 345–353.

Potts, S.G., Imperatriz-Fonseca, V., Ngo, H.T., Aizen, M.A., Biesmeijer, J.C., Breeze, T.D., Dicks, L.V., Garibaldi, L.A., Hill, R., Settele, J., Vanbergen, A.J., 2016. Safeguarding pollinators and their values to human well-being. Nature 540, 220–229.

Pywell, R.F., Heard, M.S., Woodcock, B.A., Hinsley, S., Ridding, L., Nowakowski, M., Bullock, J.M., 2015. Wildlife-friendly farming increases crop yield: evidence for ecological intensification. Proc. R. Soc. B 282, 20151740. https://doi.org/10.1098/rspb.2015.1740.

Raaijmakers, J.M., Paulitz, T.C., Steinberg, C., Alabouvette, C., Moënne-Loccoz, Y., 2009. The rhizosphere: a playground and battlefield for soilborne pathogens and beneficial microorganisms. Plant and Soil 321, 341–361. https://doi.org/10.1007/s11104-008-9568-6.

Rader, R., Reilly, J., Bartomeus, I., Winfree, R., 2013. Native bees buffer the negative impact of climate warming on honey bee pollination of watermelon crops. Glob. Chang. Biol. 19, 3103–3110.

Rader, R., Bartomeus, I., Garibaldi, L.A., Garrat, M.D.P., Howlett, B., Cunningham, S.A., Mayfield, M.M., Arthur, A.D., Andersson, G.K.S., Bommarco, R., Brittain, C., Carvalheiro, L.G., Chacoff, N.P., Entling, M.H., Foully, B., Freitas, B.M., Gemmill-Herren, B., Ghazoul, J., Griffin, S.R., Gross, C.L., Herbertsson, L., Herzog, F., Hipólito, J., Jaggar, S., Jauker, F., Klein, A.M., Kleijn, D., Krishnan, S., Lemos, C.Q., Lindström, S.A.M., Mandelik, Y., Monteiro, V.M., Nelson, W., Nilsson, L., Pattemore, D.E., Pereira, N.O., Pisanty, G., Potts, S.G., Reemer, M., Rundlöf, M., Sheffield, C.S., Scheper, J., Schüepp, C., Smith, H.G., Stanley, D.A., Stout, J.C., Szentgyörgyi, H., Taki, H., Vergara, C.H., Viana, B.F., Woyciechowski, M., 2016. Non-bee insects are important contributors to global crop pollination. Proc. Natl. Acad. Sci. U. S. A. 113, 146–151.

Redlich, S., Martin, E.A., Steffan-Dewenter, I., 2018. Landscape-level crop diversity benefits biological pest control. J. Appl. Ecol. 55, 2419–2428. https://doi.org/10.1111/1365-2664.13126.

Ricketts, T.H., Regetz, J., Steffan-Dewenter, I., Cunningham, S.A., Kremen, C., Bogdanski, A., Gemmill-Herren, B., Greenleaf, S.S., Klein, A.M., Mayfield, M.M., Morandin, L.A., Ochieng, A., Potts, S.G., Viana, B.F., 2008. Landscape effects on crop pollination services: are there general patterns? Ecol. Lett. 11, 499–515. https://doi.org/10.1111/j.1461-0248.2008.01157.x.

Romo, C.M., Tylianakis, J.M., 2013. Elevated temperature and drought interact to reduce parasitoid effectiveness in suppressing hosts. PLoS One 8, e58136.

Roslin, T., Hardwick, B., Novotny, V., Petry, W.K., Andrew, N.R., Asmus, A., Barrio, I.C., Basset, Y., Boesing, A.L., Bonebrake, T.C., Cameron, E.K., Dáttilo, W., Donoso, D.A., Drozd, P., Gray, C.L., Hik, D.S., Hill, S.J., Hopkins, T., Huang, S., Koane, B., Laird-Hopkins, B., Laukkanen, L., Lewis, O.T., Milne, S., Mwesige, I., Nakamura, A., Nell, C.S., Nichols, E., Prokurat, A., Sam, K., Schmidt, N.M., Slade, A., Slade, V., Suchanková, A., Teder, T., Van Nouhuys, S., Vandvik, V., Weissflog, A., Zhukovich, V., Slade, E.M., 2017. Latitudinal gradients: higher predation risk for insect prey at low latitudes and elevations. Science 356, 742–744. https://doi.org/10.1126/science.aaj1631.

Rowe, R.C., 1978. Control of Fusarium crown and root rot of greenhouse tomatoes by inhibiting recolonization of steam-disinfested soil with a captafol drench. Phytopathology 68, 1221. https://doi.org/10.1094/Phyto-68-1221.

Rundlöf, M., Andersson, G.K.S., Bommarco, R., Fries, I., Hederström, V., Herbertsson, L., Jonsson, O., Klatt, B.K., Pedersen, T.R., Yourstone, J., Smith, H.G., 2015. Seed coating with a neonicotinoid insecticide negatively affects wild bees. Nature 521, 77–80.

Rusch, A., Bommarco, R., Jonsson, M., Smith, H.G., Ekbom, B., 2013. Flow and stability of natural pest control services depend on complexity and crop rotation at the landscape scale. J. Appl. Ecol. 50, 345–354. https://doi.org/10.1111/1365-2664.12055.

Rusch, A., Chaplin-Kramer, R., Gardiner, M.M., Hawro, V., Holland, J., Landis, D., Thies, C., Tscharntke, T., Weisser, W.W., Winqvist, C., Woltz, M., Bommarco, R., 2016. Agricultural landscape simplification reduces natural pest control: a quantitative synthesis. Agric. Ecosyst. Environ. 221, 198–204. https://doi.org/10.1016/j.agee.2016.01.039.

Saez, A., Morales, C.L., Garibaldi, L.A., Aizen, M.A., 2017. Invasive bumble bees reduce nectar availability for honey bees by robbing raspberry flower buds. Basic Appl. Ecol. 19, 26–35. https://doi.org/10.1016/j.baae.2017.01.001.

Sala, O.E., Chapin III, F.S., Armesto, J.J., Berlow, E., Bloomfield, J., Dirzo, R., Huber-Sanwald, E., Huenneke, L.F., Jackson, R.B., Kinzig, A., Leemans, R., Lodge, D.M., Mooney, H.A., Oesterheld, M., Poff, N.L., Sykes, M.T., Walker, B.H., Walker, M., Wall, D.H., 2000. Global biodiversity scenarios for the year 2100. Science 287, 1770–1774. https://doi.org/10.1126/science.287.5459.1770.

Sanders, D., Thébault, E., Kehoe, R., van Veen, F.J.F., 2018. Trophic redundancy reduces vulnerability to extinction cascades. Proc. Natl. Acad. Sci. U. S. A. 115, 2419–2424. https://doi.org/10.1073/pnas.1716825115.

Sasaki, T., Furukawa, T., Iwasaki, Y., Seto, M., Mori, A.S., 2015. Perspectives for ecosystem management based on ecosystem resilience and ecological thresholds against multiple and stochastic disturbances. Ecol. Indic. 57, 395–408.

Scheffer, M., Bascompte, J., Brock, W.A., Brovkin, V., Carpenter, S.R., Dakos, V., Held, H., van Nes, E.H., Rietkerk, M., Sugihara, G., 2009. Early-warning signals for critical transitions. Nature 461, 53–59. https://doi.org/10.1038/nature08227.

Scheffer, M., Carpenter, S.R., Dakos, V., van Nes, E.H., 2015. Generic indicators of ecological resilience: inferring the chance of a critical transition. Annu. Rev. Ecol. Evol. Syst. 46, 145–167. https://doi.org/10.1146/annurev-ecolsys-112414-054242.

Schellhorn, N.A., Bianchi, F., Hsu, C.L., 2014. Movement of entomophagous arthropods in agricultural landscapes: links to pest suppression. Annu. Rev. Entomol. 59, 559–581.

Schellhorn, N.A., Gagic, V., Bommarco, R., 2015. Time will tell: resource continuity bolsters ecosystem services. Trends Ecol. Evol. 30, 524–530.

Schlatter, D., Kinkel, L., Thomashow, L., Weller, D., Paulitz, T., 2017. Disease suppressive soils: new insights from the soil microbiome. Phytopathology 107, 1284–1297.

Siegel-Hertz, K., Edel-Hermann, V., Chapelle, E., Terrat, S., Raaijmakers, J.M., Steinberg, C., 2018. Comparative microbiome analysis of a Fusarium wilt suppressive soil and a Fusarium wilt conducive soil from the Châteaurenard region. Front. Microbiol. 9, 568. https://doi.org/10.3389/fmicb.2018.00568.

Smolinska, U., Morra, M.J., Knudsen, G.R., James, R.L., 2003. Isothiocyanates produced by Brassicaceae species as inhibitors of Fusarium oxysporum. Plant Dis. 87, 407–412. https://doi.org/10.1094/PDIS.2003.87.4.407.

Standish, R.J., Hobbs, R.J., Mayfield, M.M., Bestelmeyer, B.T., Suding, K.N., Battaglia, L.L., Eviner, V., Hawkes, C.V., Temperton, V.M., Cramer, V.A., 2014. Resilience in ecology: abstraction, distraction, or where the action is? Biol. Conserv. 177, 43–51.

Stapel, J.O., Cortesero, A.M., Lewis, W.J., 2000. Disruptive sublethal effects of insecticides on biological control: altered foraging ability and life span of a parasitoid after feeding on extrafloral nectar of cotton treated with systemic insecticides. Biol. Control 17, 243–249. https://doi.org/10.1006/bcon.1999.0795.

Staudacher, K., Rennstam Rubbmark, O., Birkhofer, K., Malsher, G., Sint, D., Jonsson, M., Traugott, M., 2018. Habitat heterogeneity induces rapid changes in the feeding behaviour of generalist arthropod predators. Funct. Ecol. 32, 809–819. https://doi.org/10.1111/1365-2435.13028.

Stavert, J.R., Pattemore, D.E., Bartomeus, I., Gaskett, A.C., Beggs, J.R., 2018. Exotic flies maintain pollination services as native pollinators decline with agricultural expansion. J. Appl. Ecol. 55, 1737–1746. https://doi.org/10.1111/1365-2664.13103.

Steffan-Dewenter, I., Schiele, S., 2008. Do resources or natural enemies drive bee population dynamics in fragmented habitats? Ecology 89, 1375–1387.

Steffan-Dewenter, I., Münzenberg, U., Bürger, C., Thies, C., Tscharntke, T., 2002. Scale-dependent effects of landscape context on three pollinator guilds. Ecology 83, 1421–1432.

Steffen, W., Rockström, J., Richardson, K., Lenton, T.M., Folke, C., Liverman, D., Summerhayes, C.P., Barnosky, A.D., Cornell, S.E., Crucifix, M., 2018. Trajectories of the earth system in the anthropocene. Proc. Natl. Acad. Sci. U. S. A. 115, 8252–8259.

Sterling, E.J., Filardi, C., Toomey, A., Sigouin, A., Betley, E., Gazit, N., Newell, J., Albert, S., Alvira, D., Bergamini, N., 2017. Biocultural approaches to well-being and sustainability indicators across scales. Nat. Ecol. Evol. 1, 1798.

Stres, B., Philippot, L., Faganeli, J., Tiedje, J.M., 2010. Frequent freeze–thaw cycles yield diminished yet resistant and responsive microbial communities in two temperate soils: a laboratory experiment. FEMS Microbiol. Ecol. 74, 323–335. https://doi.org/10.1111/j.1574-6941.2010.00951.x.

Suding, K.N., Lavorel, S., Chapin, F.S., Cornelissen, J.H.C., Díaz, S., Garnier, E., Goldberg, D., Hooper, D.U., Jackson, S.T., Navas, M.-L., 2008. Scaling environmental change through the community-level: a trait-based response-and-effect framework for plants. Glob. Chang. Biol. 14, 1125–1140. https://doi.org/10.1111/j.1365-2486.2008.01557.x.

Tamburini, G., De Simone, S., Sigura, M., Boscutti, F., Marini, L., 2016. Conservation tillage mitigates the negative effect of landscape simplification on biological control. J. Appl. Ecol. 53, 233–241. https://doi.org/10.1111/1365-2664.12544.

Termorshuizen, A.J., Jeger, M.J., 2008. Strategies of soilborne plant pathogenic fungi in relation to disease suppression. Fungal Ecol. 1, 108–114. https://doi.org/10.1016/j.funeco.2008.10.006.

Termorshuizen, A.J., van Rijn, E., van der Gaag, D.J., Alabouvette, C., Chen, Y., Lagerlöf, J., Malandrakis, A.A., Paplomatas, E.J., Rämert, B., Ryckeboer, J., Steinberg, C., Zmora-Nahum, S., 2006. Suppressiveness of 18 composts against 7 pathosystems: variability in pathogen response. Soil Biol. Biochem. 38, 2461–2477. https://doi.org/10.1016/j.soilbio.2006.03.002.

Thébault, E., Fontaine, C., 2010. Stability of ecological communities and the architecture of mutualistic and trophic networks. Science 329, 853–856. https://doi.org/10.1126/science.1188321.

Thibaut, L.M., Connolly, S.R., 2013. Understanding diversity-stability relationships: towards a unified model of portfolio effects. Ecol. Lett. 16, 140–150. https://doi.org/10.1111/ele.12019.

Thies, C., Tscharntke, T., 1999. Landscape structure and biological control in agroecosystems. Science 285, 893–895.

Thies, C., Haenke, S., Scherber, C., Bengtsson, J., Bommarco, R., Clement, L.W., Ceryngier, P., Dennis, C., Emmerson, M., Gagic, V., Hawro, V., Liira, J., Weisser, W.W., Winqvist, C., Tscharntke, T., 2011. The relationship between agricultural intensification and biological control: experimental tests across Europe. Ecol. Appl. 21, 2187–2196. https://doi.org/10.1890/10-0929.1.

Thomson, L.J., Macfadyen, S., Hoffmann, A.A., 2010. Predicting the effects of climate change on natural enemies of agricultural pests. Biol. Control 52, 296–306. https://doi.org/10.1016/J.BIOCONTROL.2009.01.022.

Thorbek, P., Bilde, T., 2004. Reduced numbers of generalist arthropod predators after crop management. J. Appl. Ecol. 41, 526–538. https://doi.org/10.1111/j.0021-8901.2004.00913.x.

Thrush, S.F., Hewitt, J.E., Dayton, P.K., Coco, G., Lohrer, A.M., Norkko, A., Norkko, J., Chiantore, M., 2009. Forecasting the limits of resilience: integrating empirical research with theory. Proc. R. Soc. Lond. B Biol. Sci. 276, 3209–3217. https://doi.org/10.1098/rspb.2009.0661.

Trivedi, P., Delgado-Baquerizo, M., Trivedi, C., Hamonts, K., Anderson, I.C., Singh, B.K., 2017. Keystone microbial taxa regulate the invasion of a fungal pathogen in agro-ecosystems. Soil Biol. Biochem. 111, 10–14. https://doi.org/10.1016/j.soilbio.2017.03.013.

Tscharntke, T., Bommarco, R., Clough, Y., Crist, T.O., Kleijn, D., Rand, T.A., Tylianakis, J.M., van Nouhuys, S., Vidal, S., 2008a. Reprint of "conservation biological control and enemy diversity on a landscape scale" [Biol. Control 43 (2007) 294–309]. Biol. Control 45, 238–253.

Tscharntke, T., Sekercioglu, C.H., Dietsch, T.V., Sodhi, N.S., Hoehn, P., Tylianakis, J.M., 2008b. Landscape constraints on functional diversity of birds and insects in tropical agroecosystems. Ecology 89, 944–951. https://doi.org/10.1890/07-0455.1.

Tscharntke, T., Tylianakis, J.M., Rand, T.A., Didham, R.K., Fahrig, L., Batary, P., Bengtsson, J., Clough, Y., Crist, T.O., Dormann, C.F., et al., 2012. Landscape moderation of biodiversity patterns and processes-eight hypotheses. Biol. Rev. 87, 661–685.

Tscharntke, T., Karp, D.S., Chaplin-Kramer, R., Batáry, P., DeClerck, F., Gratton, C., Hunt, L., Ives, A., Jonsson, M., Larsen, A., Martin, E.A., Martínez-Salinas, A., Meehan, T.D., O'Rourke, M., Poveda, K., Rosenheim, J.A., Rusch, A., Schellhorn, N., Wanger, T.C., Wratten, S., Zhang, W., 2016. When natural habitat fails to enhance biological pest control—five hypotheses. Biol. Conserv. 204 (Pt. B), 449–458. https://doi.org/10.1016/j.biocon.2016.10.001.

Tuck, S.L., Winqvist, C., Mota, F., Ahnström, J., Turnbull, L.A., Bengtsson, J., 2014. Land-use intensity and the effects of organic farming on biodiversity: a hierarchical meta-analysis. J. Appl. Ecol. 51, 746–755. https://doi.org/10.1111/1365-2664.12219.

Tylianakis, J.M., Morris, R.J., 2017. Ecological networks across environmental gradients. Annu. Rev. Ecol. Evol. Syst. 48, 25–48. https://doi.org/10.1146/annurev-ecolsys-110316-022821.

Tylianakis, J.M., Tscharntke, T., Klein, A.M., 2006. Diversity, ecosystem function, and stability of parasitoid-host interactions across a tropical habitat gradient. Ecology 87, 3047–3057. https://doi.org/10.1890/0012-9658(2006)87[3047:DEFASO]2.0.CO;2.

Tylianakis, J.M., Didham, R.K., Bascompte, J., Wardle, D.A., 2008. Global change and species interactions in terrestrial ecosystems. Ecol. Lett. 11, 1351–1363. https://doi.org/10.1111/j.1461-0248.2008.01250.x.

Valone, T.J., Barber, N.A., 2008. An empirical evaluation of the insurance hypothesis in diversity–stability models. Ecology 89, 522–531. https://doi.org/10.1890/07-0153.1.

van de Leemput, I.A., Dakos, V., Scheffer, M., van Nes, E.H., 2018. Slow recovery from local disturbances as an indicator for loss of ecosystem resilience. Ecosystems 21, 141–152. https://doi.org/10.1007/s10021-017-0154-8.

van Elsas, J.D., Chiurazzi, M., Mallon, C.A., Elhottova, D., Kristufek, V., Salles, J.F., 2012. Microbial diversity determines the invasion of soil by a bacterial pathogen. Proc. Natl. Acad. Sci. U. S. A. 109, 1159–1164. https://doi.org/10.1073/pnas.1109326109.

van Vliet, J., de Groot, H.L.F., Rietveld, P., Verburg, P.H., 2015. Manifestations and underlying drivers of agricultural land use change in Europe. Landsc. Urban Plan. 133, 24–36. https://doi.org/10.1016/j.landurbplan.2014.09.001.

Vandermeer, J., 2011. The inevitability of surprise in agroecosystems. Ecol. Complex 8, 377–382. Special Section: Complexity of Coupled Human and Natural Systems. https://doi.org/10.1016/j.ecocom.2011.10.001.

Vasseur, C., Joannon, A., Aviron, S., Burel, F., Meynard, J.-M., Baudry, J., 2013. The cropping systems mosaic: how does the hidden heterogeneity of agricultural landscapes drive arthropod populations? Agric. Ecosyst. Environ. 166, 3–14. https://doi.org/10.1016/j.agee.2012.08.013.

Veraart, A.J., Faassen, E.J., Dakos, V., van Nes, E.H., Lürling, M., Scheffer, M., 2012. Recovery rates reflect distance to a tipping point in a living system. Nature 481, 357–359. https://doi.org/10.1038/nature10723.

Walker, B., Holling, C.S., Carpenter, S., Kinzig, A., 2004. Resilience, adaptability and transformability in social–ecological systems. Ecol. Soc. 9, 5. https://doi.org/10.5751/ES-00650-090205.

Wang, Q., Ma, Y., Wang, G., Gu, Z., Sun, D., An, X., Chang, Z., 2014. Integration of biofumigation with antagonistic microorganism can control *Phytophthora* blight of pepper plants by regulating soil bacterial community structure. Eur. J. Soil Biol. 61, 58–67. https://doi.org/10.1016/j.ejsobi.2013.12.004.

Weerakoon, D.M.N., Reardon, C.L., Paulitz, T.C., Izzo, A.D., Mazzola, M., 2012. Long-term suppression of *Pythium abappressorium* induced by *Brassica juncea* seed meal amendment is biologically mediated. Soil Biol. Biochem. 51, 44–52. https://doi.org/10.1016/j.soilbio.2012.03.027.

Wei, Z., Yang, T., Friman, V.-P., Xu, Y., Shen, Q., Jousset, A., 2015. Trophic network architecture of root-associated bacterial communities determines pathogen invasion and plant health. Nat. Commun. 6, 8413. https://doi.org/10.1038/ncomms9413.

Wei, F., Passey, T., Xu, X., 2016. Amplicon-based metabarcoding reveals temporal response of soil microbial community to fumigation-derived products. Appl. Soil Ecol. 103, 83–92. https://doi.org/10.1016/j.apsoil.2016.03.009.

Weibull, A.C., Östman, Ö., Granqvist, Å., 2003. Species richness in agroecosystems: the effect of landscape, habitat and farm management. Biodivers. Conserv. 12, 1335–1355. https://doi.org/10.1023/A:1023617117780.

Weise, H., Auge, H., Baessler, C., Baerlund, I., Bennett, E.M., Berger, U., Bohn, F., Bonn, A., Borchardt, D., Brand, F., Chatzinotas, A., Corstanje, R., Laender, F.D., Dietrich, P., Dunker, S., Durka, W., Fazey, I., Groeneveld, J., Guilbaud, C.S.E., Harms, H., Harpole, S., Harris, J.A., Jax, K., Jeltsch, F., Johst, K., Joshi, J., Klotz, S., Kuehn, I., Kuhlicke, C., Mueller, B., Radchuk, V., Reuter, H., Rinke, K., Schmitt-Jansen, M., Seppelt, R., Singer, A.S., Standish, R.J., Thulke, H.-H., Tietjen, B., Weitere, M., Wirth, C., Wolf, C., Grimm, V., 2019. Resilience trinity: safeguarding ecosystem services across three different time horizons and decision contexts. bioRxiv, 549873. https://doi.org/10.1101/549873.

Weller, D.M., Raaijmakers, J.M., Gardener, B.B.M., Thomashow, L.S., 2002. Microbial populations responsible for specific soil suppressiveness to plant pathogens. Annu. Rev. Phytopathol. 40, 309–348. https://doi.org/10.1146/annurev.phyto.40.030402. 110010.

Wheeler, T., von Braun, J., 2013. Climate change impacts on global food security. Science 341, 508–513. https://doi.org/10.1126/science.1239402.

Williams, P.H., Osborne, J.L., 2009. Bumblebee vulnerability and conservation world-wide. Apidologie 40, 367–387.

Williams, N.M., Crone, E.E., T'ai, H.R., Minckley, R.L., Packer, L., Potts, S.G., 2010. Ecological and life-history traits predict bee species responses to environmental disturbances. Biol. Conserv. 143, 2280–2291. https://doi.org/10.1016/j.biocon. 2010.03.024.

Williams, N.M., Ward, K.L., Pope, N., Isaacs, R., Wilson, J., May, E.A., Ellis, J., Daniels, J., Pence, A., Ullmann, K., Peters, J., 2015. Native wildflower plantings support wild bee abundance and diversity in agricultural landscapes across the United States. Ecol. Appl. 25, 2119–2131.

With, K.A., Pavuk, D.M., Worchuck, J.L., Oates, R.K., Fisher, J.L., 2002. Threshold effects of landscape structure on biological control in agroecosystems. Ecol. Appl. 12, 52–65. https://doi.org/10.1890/1051-0761.

Woodcock, B.A., Redhead, J., Vanbergen, A.J., 2010. Impact of habitat type and landscape structure on biomass, species richness and functional diversity of ground beetles. Agric. Ecosyst. Environ. 139, 181–186.

Woodcock, B.A., Isaac, N.J., Bullock, J.M., Roy, D.B., Garthwaite, D.G., Crowe, A., Pywell, R.F., 2016. Impacts of neonicotinoid use on long-term population changes in wild bees in England. Nat. Commun. 7, 12459.

Yachi, S., Loreau, M., 1999. Biodiversity and ecosystem productivity in a fluctuating environment: the insurance hypothesis. Proc. Natl. Acad. Sci. U. S. A. 96, 1463–1468.

Yohalem, D., Passey, T., 2011. Amendment of soils with fresh and post-extraction lavender (*Lavandula angustifolia*) and lavandin (*Lavandula × intermedia*) reduce inoculum of *Verticillium dahliae* and inhibit wilt in strawberry. Appl. Soil Ecol. 49, 187–196. https://doi.org/10.1016/j.apsoil.2011.05.006.

Yulianti, T., Sivasithamparam, K., Turner, D.W., 2007. Saprophytic and pathogenic behaviour of R. solani AG2-1 (ZG-5) in a soil amended with *Diplotaxis tenuifolia* or *Brassica nigra* manures and incubated at different temperatures and soil water content. Plant and Soil 294, 277–289. https://doi.org/10.1007/s11104-007-9254-0.

Further reading

González-Varo, J.P., Geldmann, J., 2018. Response to comments of "conserving honey bees does not help wildlife" Science 360, 390.

CHAPTER FOUR

Modelling land use dynamics in socio-ecological systems: A case study in the UK uplands

Mette Termansen[a,*], Daniel S. Chapman[b], Claire H. Quinn[c], Evan D.G. Fraser[d], Nanlin Jin[e], Nesha Beharry-Borg[f], Klaus Hubacek[g,h]

[a]Food and Resource Economics, University of Copenhagen, Frederiksberg C, Denmark
[b]Biological and Environmental Sciences, Faculty of Natural Science, University of Stirling, Stirling, United Kingdom
[c]Sustainability Research Institute, School of Earth and Environment, University of Leeds, Leeds, United Kingdom
[d]Department of Geography, Environment, and Geomatics, University of Guelph, Guelph, ON, Canada
[e]Department of Computer and Information Sciences, Northumbria University, Newcastle upon Tyne, United Kingdom
[f]Centrascape Mid Centre Mall Compound, Chaguanas, Trinidad
[g]Center for Energy and Environmental Sciences (IVEM), Energy and Sustainability Research Institute Groningen (ESRIG), University of Groningen, Groningen, The Netherlands
[h]International Institute for Applied Systems Analysis, Laxenburg, Austria
*Corresponding author: e-mail address: mt@ifro.ku.dk

Contents

Advances in Ecological Research, Volume 60
ISSN 0065-2504
https://doi.org/10.1016/bs.aecr.2019.03.002

Abstract

It is well-recognised that to achieve long-term sustainable and resilient land management we need to understand the coupled dynamics of social and ecological systems. Land use change scenarios will often aim to understand (i) the behaviours of land management, influenced by direct and indirect drivers, (ii) the resulting changes in land use and (iii) the environmental implications of these changes. While the literature in this field is extensive, approaches to parameterise coupled systems through integration of empirical social science based models and ecology based models still need further development. We propose an approach to land use dynamics modelling based on the integration of behavioural models derived from choice experiments and spatially explicit systems dynamics modelling. This involves the specification of a choice model to parameterise land use behaviour and the integration with a spatial habitat succession model.

We test this approach in an upland socio-ecological system in the United Kingdom. We conduct a choice experiment with land managers in the Peak District National Park. The elicited preferences form the basis for a behavioural model, which is integrated with a habitat succession model to predict the landscape level vegetation impacts. The integrated model allows us to create projections of how land use may change in the future under different environmental and policy scenarios, and the impact this may have on landscape vegetation patterns. We illustrate this by showing future projection of landscape changes related to hypothetical changes to EU level agricultural management incentives.

The advantages of this approach are (i) the approach takes into account potential environmental and management feedbacks, an aspect often ignored in choice modelling, (ii) the behavioural rules are revealed from actual and hypothetical choice data, which allow the research to test the empirical evidence for various determinants of choice, (iii) the behavioural choice models generate probabilities of alternative behaviours which make them ideally suited for integration with simulation models.

The paper concludes that the modelling approach offers a promising route for linking socio-economic and ecological features of socio-ecological systems. Furthermore, our proposed approach allows testing of the underlying socio-economic and environmental drivers and their interaction in real environmental systems.

1. Introduction

An overarching aim of land use change modelling in environmental research is first to understand how the spatial and temporal patterns of land managers' behaviours shape agricultural or forest systems (Agarwal et al., 2002). Second, how this leads to diverse environmental impacts, e.g., on biodiversity conservation (Drechsler et al., 2007), on water quality (Nainggolan et al., 2018) and on climate services (Prestele et al., 2017; Zandersen et al., 2016). Furthermore, models might also be designed to inform policy debates by providing a framework to anticipate and test how land use patterns may change under different hypothetical future scenarios (e.g. Tieskens et al., 2017; Wolff et al., 2018). Several interesting approaches have emerged to address different aspects of such research questions. In this paper, we focus on two of these, dynamic systems models and in particular agent based simulation models (ABMs) (An, 2012) and statistically based approaches using choice models (CMs) (Train, 2009).

In most cases, both ABMs and CMs have been used independently to understand the rationale and implications of land-use decisions. Several key developments in the application of ABMs to model land-use patterns have emerged over a long period of time (see e.g. Agarwal et al., 2002; Bousquet et al., 1998; Huber et al., 2018; Veldkamp and Verburg, 2004). These contributions have addressed interactions among individuals to simulate land-use decisions of individual agents. For example, Barreteau and Bousquet (2000) used an ABM to understand the influence of existing social networks on the viability of an irrigated farming system in the Senegal River Valley. However, it is acknowledged that it is still challenging to move beyond hypothetical landscapes, and parameterise and validate the interactions needed to model real landscapes (Huber et al., 2018). Never the less, a few studies have attempted to develop more realistic systems (see e.g. Evans and Kelley, 2004; Tieskens et al., 2017) and accounting for the role of decisions of heterogonous agents in land-use changes (Huigen, 2004). A similar line of land-use related contributions has sought to determine how new agricultural practices are adopted by a population of farmers to better understand diffusion of technology (Balmann, 1997, Berger, 2001, Polhill et al., 2001; Schreinemachers and Berger, 2011).

Another research approach used to explain or predict land managers' decisions is choice experiment (CE) designed to parameterise choice models (CMs). These have been used to understand and predict farmers' land-use decisions under a variety of hypothetical scenarios using data based on individuals' responses collected using questionnaires. One such example is the study by Baltas and Korka (2002) that used a nested discrete choice model to analyse land-use allocation under risk. A number of choice modelling applications have investigated the determinants explaining farmers' participation in environmental programs that would eventually lead to changes in land-use (Aslam et al., 2017; Cooper and Keim, 1996; Lynch et al., 2002; Peterson et al., 2007; Shaikh et al., 2005; Vanslembrouck et al., 2002; Vedel et al., 2015). These studies have determined the probability of farmers' participation in hypothetical agricultural improvement programs and estimated their willingness to accept (WTA) compensation for participation in these programs.

The two approaches outlined above, therefore, have complementary advantages for advancing how we integrate and model coupled social-ecological systems. The ABM has the advantage of being a dynamic framework, incorporating feedback mechanisms that allow behaviour to be investigated in coupled systems. However, the specification of human behaviour and its interaction with the natural resource is often highly abstract and few models in the literature are parameterised or validated using real spatially specific data. On the other hand, CMs have been developed to parameterise behavioural models and capture heterogeneity between agent types (Boxall and Adamowicz, 2002; Milon and Scrogin, 2006; Train, 2016). However, CEs have usually not been developed to analyse emergent system properties as agent. They are commonly assumed to be acting independently (perhaps with the exception of social interaction models e.g. Brock and Durlauf, 2006) and the dynamic environmental feedbacks are usually ignored. CMs do, however, have the potential to provide a theoretical and empirical foundation for the link between changing environments and the consequential responses of land managers. The combination of these two approaches to achieve a dynamically integrated model with empirically estimated behavioural rules will, therefore, offer an advancement to modelling of land use dynamics.

The combined use of ABMs and CMs appears to be rare in the literature on the environmental implications of land use decisions. One exception is the study combining a catchment scale hydrological simulation model with a behavioural model of homeowners lawn watering behaviour to improve prediction of water demand (Conrad and Yates, 2018). However, there are studies that integrate the two approaches in other fields. Notably, in the field of transportation studies aiming to link individual road user behaviour with aggregated traffic system dynamics (Dia, 2001; Takama and Preston, 2008). Takama and Preston investigate the effect of a road user charging scheme for visitors to the Upper Derwent Valley in the Peak District National Park. The authors use an ABM to incorporate an interaction term that accounts for congestion thereby enhancing the discrete choice model. Dia (2001) uses a discrete choice survey to obtain each driver's preferences and characteristics and each driver is then modelled as an agent. Agents interact with their environment and with other agents in the system by receiving and reacting to real-time traffic information. In the outdoor recreation context, Hunt et al. (2007) developed a landscape fisheries model by using information from a revealed preference choice model and agent based models. Revealed preference data from each angler's trip and data on available fishing sites were used to parameterise an ABM for recreational fishing in Northern Ontario.

Building on the previous work, we propose an integrated ABM and CM by using a choice experiment to parameterise land managers behaviour in an agent based simulation model. We apply this approach in the upland ecosystems of the United Kingdom where land management is dominated by sheep and game bird production. To characterise agent behaviour we conduct a choice experiment with land managers in the Peak District National Park. The choice experiment was designed to reveal preferences for alternative production strategies under alternative scenarios of future change and to investigate the dependence of such choices on the changing environmental characteristics across the individuals' land management units. The choice data form the basis for a behavioural model, which is integrated with a habitat succession model, to form an integrated model of landscape habitat dynamics. The integrated model allows us to evaluate future changes in land use and the way these changes impact the environment under different environmental and policy scenarios. The data stems from an earlier study by Chapman et al. (2009). However in this study we take the analysis further to improve understanding of the underlying drivers of the dynamics of the joint production system. Furthermore, we illustrate how the coupled model can be used to analyse potential impacts of agri-environmental policy reforms. We illustrate this by showing future projections of landscape changes related to hypothetical changes to the level of the current EU agricultural subsidies. This scheme is also known as the single farm payment (SFP), which is an area-based payment. The payment schemes are currently being revised (COM, 2017) and it is therefore interesting to understand the impacts of proposed changes.

2. Methodology and data

2.1 The study site

The Peak District National park was established in 1951 and was the UK's first National Park. Its central location within easy reach of approximately 48% of England's population makes it one of the world's most visited national parks, with over 22 million visitor days a year (Peak District National Park, 2004). In addition to visitors, 38,000 people live within the park boundaries (Office for National Statistics, 2003) and 12% of them are employed in agriculture. However, agriculture is financially marginal, most farmers rely on government subsidies and 93% of the national park is designated as a "Less Favoured Area" (European Commission Directive 75/268) (Dougill et al., 2006).

This study focused on the Dark Peak area in the north of the national park. The Dark Peak covers more than half of the national park, is predominantly on acidic peat soils over millstone grit and supports over 17,000 ha of heather moorland and almost 16,000 ha of blanket bog (Sustainable Uplands and Moors for the Future, 2007) (Fig. 1). Both of these habitats are recognised as internationally important. Both heather moorland and blanket bog are recognised as key biodiversity habitats (UK Biodiversity Steering Group, 1995), have been designated as Sites of Special Scientific Interest (SSSI) (English Nature, 2003) and are listed in the EU's Habitats Directive (92/43/EEC) as requiring special conservation measures as Special Areas of

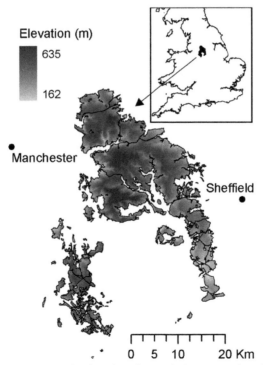

Fig. 1 Map of the open moorland within the Peak District and its location in Great Britain. To preserve confidentiality we are unable to show the exact location of the modelled management units.

Conservation (SAC) and Special Protection Areas (SPA). Extensive sheep farming and grouse (*Lagopus lagopus scoticus*) moor management form the dominant land uses.

Sheep production has traditionally been subsidised by the European Union's Common Agricultural Policy (CAP), while grouse production has not. However, proposed changes to the single farm payment, SFP, scheme may shift the incentives to managing different types of production systems (COM, 2017). The key habitat management tool for grouse moors is rotational burning of heather to provide a mosaic of different age heather stands to maximise territories for grouse (Watson and Miller, 1976).

3. Conceptual model overview

We describe the upland system as a joint production system of sheep and grouse in which the farmer chooses stocking density throughout the year, whether or not to burn and the burning schedule. We focused on sheep and grouse as they have historically been, and still is, important for both the upland economy and the characteristic upland vegetation and landscape. By choosing the stocking densities and burning schedule the farmer impacts the costs of the production system as they arise from maintenance of the sheep flock, in particular over wintering, and the labour costs involved in heather moorland management. We refer to the combination of summer and winter sheep stocking density and the burning schedule as the "moorland management regime", and the individual management characteristics (e.g. summer grazing density) as the management attributes. It is through the choice of moorland management regime that farmers impact their own livelihood, the broader upland economy and the upland environment, both the vegetation itself and other environmental variables such as water quality, carbon storage, and biodiversity. The interdependence between the sheep and grouse relates to the abundance of the preferred habitat of sheep relative to grouse, which in turn is partially determined by the choice of management regime.

The aim is to model the choices that farmers make both under current conditions and under hypothetical changes to the current situation to identify the preferable management regimes, from a farmers point of view, under varying environmental and economic conditions. It is important to note that we do not assume that farmers aim to maximise production or income from their land, and we merely characterise the importance of environmental and economic factors for variation in land management choices across the landscape. In other words, we test whether we can estimate an attractiveness (or "utility" using economic language) function of land management regimes by a combination of variables defining the moorland management regime, the environmental conditions of the land and the socio-economic conditions of the farmer. If actual and hypothetical choices are described well by the estimated model, this will help to understand better the spatial variation in management choices across the uplands and across upland farmers.

In this study, we test four basic hypotheses/relationships: (i) Land managers will display their relative preference for grouse production by choosing land management regimes favouring diverse heather age distributions; (ii) optimal stocking density and optimal burning regime (as perceived by the land managers) will vary across the landscape reflecting the variation in the carrying capacity for sheep and grouse and the land manager's desire to keep management costs low by reducing the number of wintering sheep and the area of burning; (iii) the extent to which management choices will vary with socio-economic characteristics of the land manager and the condition under which the land is managed, i.e., ownership or tenancy agreements; (iv) the robustness of the perceived optimal moorland management to changes in the level of the single farm payment, SFP.

In the model, land managers choose the favoured management regime given the characteristics of the plot of land. This results in a change in vegetation cover (of predominantly heather and grass), which in turn changes the perceived optimal strategy of the land manager. We investigate the dynamics of the coupled system of land management choices and habitat response. We compare scenarios by testing the differences between simulation outputs when changing amount of the single farm payment. Changing the mean temperature uniformly across the landscape simulates the impacts that climate change may have.

4. Data

4.1 Data collected to characterise the land managers' behaviour

Face-to-face interviews were carried out with land managers in the Dark Peak area in 2006 and 2007. Potential respondents were identified using two methods. An initial contact list was generated from meetings with key stakeholders in the Peak District National Park. This allowed the interviews to be targeted only to land managers within the Dark Peak area and reduced the number of redundant contacts. Additional respondents were then identified using the snowball method (Bailey, 1982) where those taking part were then asked to identify others. It is important to note that land managers are often responsible for managing several areas with different management regimes. Therefore, each contact was asked separately about land management on all the different "land management units" under their control. We obtained complete observations of environmental and behavioural data for a total of 71 individual management units, representing 40% of the Dark Peak area. This was accomplished in 25 face-to-face-interviews.

Data were collected during the interviews using a standardised questionnaire based on both structured open-ended and closed-format questions. The questionnaire was piloted with key stakeholders before being used with respondents.

The first part of the questionnaire focused on identifying the type of land ownership (private versus tenancy) and the dominant land management activity carried out by the respondent (grouse moor management versus sheep farming). The extent of the land they managed was also identified using Ordnance Survey maps. Land was categorised into unique management units, reflecting heather moorland, rough grazing and improved pasture and the geographical boundaries between management units were recorded. The second part of the questionnaire was carried out for each of the identified management units and used a choice card for respondents to characterise their management. Respondents were first asked to characterise their current management strategies according to the variables on the card (Table 1) and this was recorded in the questionnaire. From this baseline respondents were then asked how their management would change under different hypothetical scenarios. In this paper we focus on the scenario related to changes in the single farm payment. In this scenario we asked land managers "What if the current business environment stayed the same but the payments you currently receive from the single farm subsidy scheme changed so that you received £X per hectare, would your management change?" X was drawn randomly from the following set: 10, 20, 30, 40, and 60. The number of repetitions (variations in X) of this experiment varied between interviews, depending on the number of management units, to limit the total amount of time spent on the questionnaire by each respondent. The maximum number of repetitions was four and the minimum two. This resulted in a total data set of 200 observations (Table 1).

Table 1 Data tested for the specification of the behavioural model.

Variable	Description	Source
Land Holding, LH	Geographical mapping of individual land holdings	Interviews
Management Unit, MU	Geographical mapping of areas with uniform management strategy. Digitised in ArcGIS	Interviews
Tenancy, T	Owned by land manager, land manager tenant on the land, land manager employed	Interviews
Dominant land use DOM	Land predominantly used for grazing, land predominantly used for grouse shooting, Land predominantly used for other purpose	Interviews
Grazing time, GT	The temporal extent of the grazing; no grazing, spring and summer, all year with 25% reduction in winter, all year	Interviews Choice card
Stocking Rate, SR	Stocking Rate when sheep are grassing the land; no grazing, 0.5 sheep/ha, 1 sheep/ha, 2 sheep/ha, 3 sheep/ha	Interviews Choice card
Stocking Density, SD	Average summer stocking density, calculated by combining GT and SR	Derived
Burning Frequency, BF	Frequency of burning in a management unit; every year, every other year, every third year, every fifth year, never	Interviews Choice card
Burning Proportion, BP	Proportion burned when management unit is burned; 0.5%, 10%, 15% and 20%	Interviews Choice card
Shannon Diversity Index, SH	Shannon-Weaver's Diversity Index of dwarf shrubs age distribution. Calculated from each management option (combination of BF and BP) in a simple dwarf shrub succession model on a 100×100 grid of dwarf shrub	Derived
Single Farm Payment, SFP	Actual and hypothetical area payment independent of farming intensity; 0, 10, 20, 40, 60 £/ha	Interviews
SFP_L	Dummy variable for SFP, $SFP_L = 1$ if $SFP < L$, otherwise $SPF_L = 0$	Derived
SFP_M	Dummy variable for SFP, $SFP_M = 1$ if $L < SFP < H$, otherwise $SPF_M = 0$	Derived
SFP_H	Dummy variable for SFP, $SFP_H = 1$ if $SFP_H > H$, otherwise $SPF_H = 0$	Derived

Table 1 Data tested for the specification of the behavioural model.—cont'd

Variable	Description	Source
Temperature, Temp	Used as proxy for Elevation; to capture land carrying capacity	Met Office weather stations[a]
Grass	Proportion of gramionoids (grasses) in each MU	LCM
Heather	Proportion of dwarf shrub in each MA	LCM
D_H	Dummy variable for Heather dominant MA; $D_H = 1$ when Heather >0.5, otherwise $D_H = 0$	Derived
Burning costs, BC	Costs of burning measured as labour requirement per unit of land	Interviews
Wintering costs, WC	Costs of keeping sheep off the moors over winter	Interviews

[a]http://www.badc.nerc.ac.uk/data/ukmo-midas

4.2 Data used for the habitat succession model

Spatially-referenced environmental data used in the habitat succession model is specified in Table 2 and model formulation is outlined in the following section.

5. Model formulation and parameterisation

The model consists of two integrated components, the behavioural model and the habitat succession model. In this section, both components are described individually and then their integration and joint simulation is outlined.

5.1 Behavioural model

The underlying premise of the models we set up is that land managers choose a land management regime to maximise the utility from the joint production of grouse and sheep.

We use a random utility approach (Train, 2003) to model land managers choices between alternative management regimes. This assumes that a land manager chooses a management regime j (defined as a combination of management attributes) in an area from the set of possible regimes J with the highest expected utility from the chosen management regime. A general representative utility function of management regime j chosen by land

Table 2 Data used to run the habitat succession model.

Variables	Description	Source
LCM	Land Cover Map in 2005 grouped into dwarf shrubs, bracken, graminoid, bare peat and bare rock, 5×5 m resolution	Chapman et al. (2010)
Slope	Mean slope (degrees) derived from the NEXTmap Digital Elevation Model (5×5 m resolution)	NEXTmap[a]
Aspect	Aspect (north, south, east, west) derived from the NEXTmap Digital Elevation Model (5×5 m resolution)	NEXTmap[a]
Rainfall	Mean annual rainfall, interpolated at 100×100 m resolution	Met Office weather stations[b]
Temperature	Mean annual temperature, interpolated at 100×100 m resolution	Met Office weather stations[b]
Bedrock	Bedrock classified as either sandstone or mixed	British Geological Survey bedrock map[c]
Warming	Additional degrees Celsius added to the mean annual temperature	Scenario assumption

[a]http://www.neodc.rl.ac.uk;
[b]http://www.badc.nerc.ac.uk/data/ukmo-midas;
[c]http://www.bgs.ac.uk/products/digitalmaps/home/html

manager n is specified as v_{nj}. The utility function includes (1) regime characteristics, x_{nj}, (2) environmental characteristics, z_n, independent of the land management choices and (3) policy environment conditions, s_n, that are also independent of the choices the land manager makes. The expected utility that individual n gains from land management regime j is, therefore, comprised of a deterministic component (v_{nj}) and a random error component (ε_{nj}) specified as:

$$U_{nj} = v_{nj} + \varepsilon_{nj} \qquad (1)$$

where $v_{nj} = \boldsymbol{\beta}_1' \mathbf{x}_{nj} + \boldsymbol{\beta}_2' \mathbf{x}_{nj} z_n + \boldsymbol{\beta}_3' \mathbf{x}_{nj} s_n$. $\boldsymbol{\beta}_1$ is a preference parameter vector representing the importance of individual regime characteristics in land managers choices (i.e. parameters measuring preference for stocking density and heather diversity), $\boldsymbol{\beta}_2$ is the preference parameter vector accounting for variation across varying environmental conditions, and $\boldsymbol{\beta}_3$ accounts for the variation in preference parameters with varying policy environments.

The error component ε_{nj} is independently and identically distributed (iid) and follows a Type 1 extreme value distribution (Train, 2003).

The probability that individual n chooses regime i can be derived given the parameter vectors $\boldsymbol{\beta_1}$, $\boldsymbol{\beta_2}$, $\boldsymbol{\beta_3}$ using the multinomial logit model (MNL) specified as follows:

$$\text{Pr}_{ni} = \frac{\exp\left(v_{ni}\right)}{\sum_{j \in J} \exp\left(v_{nj}\right)} \tag{2}$$

The parameters are derived using maximum likelihood estimation using the Gauss software, version 6.0. There is no information on the subset of land management regimes considered by the land manager on each plot of land. However, all alternatives are available to the land managers and the choice set is modelled as the chosen regime and a random selection of alternatives. The advantage of this approach is the reduction in computing effort. More importantly, this approach has been shown to generate very similar relative parameter estimates to those obtained when the estimation choice set is the same as the total potential choice sets (Parsons and Kealy, 1992; Termansen et al., 2004). We test for parameter stability with increasing size of the choice set to ensure that the selected model specification is not dependent on the arbitrary choice of number of alternative regimes considered.

5.2 Habitat dynamics model

The habitat model is a spatially explicit grid based model of the dynamics of the moorland vegetation cover. Space is represented as a $100 \times 100\,\text{m}$ grid aligned to the Ordnance Survey British National Grid. The state of each grid cell is given by the cover of dwarf shrubs (predominantly heather *Calluna vulgaris*), bracken *Pteridium aquilinum*, graminoids (grasses, sedges and rushes) and bare peat. Dwarf shrubs are further divided into five growth phases based on the number of years since the last burn. The phases are newly burnt (0–2 years), pioneer (3–5 years), building (6–15 years), mature (16–25 years) and degenerate (26+ years) (Barclay-Estrup and Gimingham, 1969). It is assumed that dwarf shrubs will remain in the degenerate stage indefinitely if it is not burned, and that cells colonised by dwarf shrubs enter the pioneer stage. Aging and the choice of burning regime determine the diversity of the dwarf shrub habitat, represented in the model by the Shannon-Weaver diversity index (Shannon, 1948).

The habitat dynamics is based on the model developed by Chapman et al. (2009). Changes in habitat cover are determined by competition between dwarf shrubs, bracken and graminoids. This is mediated by grazing pressure, dwarf shrub age distribution (determined by burning) and environmental gradients. We follow the work from Palmer et al. (2004) and assume that provided grazing pressure is sufficiently low, and the environment sufficiently suitable, dwarf shrubs will dominate graminoids and bracken, and will increase in cover. We, however, modify Palmer et al.'s model for changes in dwarf shrub cover as a function of the proportion of annual productivity eaten by grazing sheep (the utilisation rate). Modifications include accounting for variations in habitat quality, the composition of competing vegetation and the dwarf growth phase following the principles described in Chapman et al. (2009). Habitat quality is determined by temperature, rainfall, topography and bedrock type. For a full specification of the vegetation dynamics refer to Chapman et al. (2009).

Managers decide how many sheep to release onto each management unit during summer and winter, but these sheep are free to move through the unit and so stocking density is therefore not uniformly distributed. We assume that sheep are distributed across the cells in proportion to the cover of their preferred forage (graminoids, including those growing within other vegetation types, following Armstrong et al., 1997a,b). Once stock density through the year has been calculated, the grazing utilisation rate of dwarf shrubs is estimated using the Hill Grazing Management Model (HGMM; Armstrong et al., 1997a,b). The HGMM is a very complex model from which we are only interested in determining of sheep grazing pressure. We have, therefore, produced a HGMM-emulator by running the model for selected variations in inputs (summer and winter stocking densities, proportions of dwarf shrub in each growth phase and temperature) and modelled the resulting grazing pressure using multiple regression. Chapman et al. (2010) give full detail of this procedure and the results obtained.

Managed burning in each management unit is implemented in each year with a probability calculated as the inverse of the burning frequency given in the land management strategy. Burning affects entire grid cells, and we assume that cells are eligible for burning if the area of dwarf shrubs exceeds a threshold of 0.3 ha, equivalent to the average size of a burning plot. The land management strategy gives the proportion of the eligible cells that are burnt each time burning occurs. These are selected as the oldest of the eligible cells. We assume that the burn is controlled and at a low temperature so that it simply resets the age of the dwarf shrub to zero. Negative impacts on dwarf shrub regeneration of high temperature burns have not been included in the model.

5.3 Simulation procedure

Management decisions are made in each year with a probability of 0.2, i.e., every 5 years on average. Decisions are made by selecting a strategy from the observed set with a probability in proportion to the values of $\exp(v_{nj})$ (Eq. 2) for all observed strategies, as determined by the choice model.

Stocking densities are distributed in each land management unit according to habitat preference of sheep and the chosen burning events are simulated by updating the state of the burned patches.

Models are initiated with the observed management strategies and vegetation cover and run to equilibrium over a 500 year transient. Outputs are then collected over the next 500 years and include the frequencies with which each management unit chooses each observed strategy. These data are used to calculate the mean values of the management strategies, such as the stocking densities, and the proportion cover of each vegetation type across the simulated landscapes.

5.4 Analysis of simulated data

Ten replicate simulations are run to establish equilibrium conditions for each combination of single farm payment level and degree of climate change. Climate change is modelled as a uniform increase in temperature by 0 to 3 °C (in 0.5 °C steps). Generalised linear models are used to analyse the components of the mean strategies outputted for each combination.

6. Results
6.1 Behavioural model specification

The selected model shows that farmers' utility from land management is dependent on stocking density, heather diversity, elevation, costs associated with sheep farming and burning and subsidy received (Table 3). The analysis did not show any evidence of dependence of the tested socio-economic characteristics of the land manager. We define low SFP as payment below £20/ha and high SFP as payments above £60 ha^{-1}, as these specifications had the best statistical fit to the observed data.

Table 3 Behavioural model specification (Choice set 100 alternative land management regimes).

Variable	Parameter	Estimate	t-statistics
Stocking Density	β_1^1	−3.828	−2.86
Socking Density × SFP$_L$	β_3^1	−0.531	−3.01
Shannon Diversity Index × Heather · D$_H$	β_2^1	4.266	4.62
Shannon Diversity Index × SFP$_H$	β_3^2	0.959	2.60
Burning Costs	β_1^2	−0.019	−3.82
Wintering Costs	β_1^3	−0.015	−1.27
Stocking Density × Temperature	β_2^2	0.5356	3.05

Heather dominated land is defined as dwarf shrub occupying >50% of the management unit. This threshold was also estimated based on statistical fit.

The probability of choosing a land management regime with a high stocking density increases with temperature reflecting the higher stocking densities in the valleys compared to the hilltops ($\beta_2^2 > 0$, $P < 0.05$). This effect is augmented by a background level propensity to choose management regimes with low sheep stocking densities ($\beta_1^1 < 0$, $P < 0.05$). It would be expected that land managers would attempt to avoid land management options that lead to high costs associated with wintering costs; however, we do not find this effect to be significant. For scenarios representing a reduction in the SFP, the probability of choosing management regimes with high stocking densities is reduced ($\beta_3^1 < 0$, $P < 0.05$). Land managers display a preference for land management regimes with higher dwarf shrub diversity on dwarf shrub dominated land ($\beta_2^1 > 0$, $P < 0.05$). High burning costs regimes are avoided, other aspect being equal ($\beta_1^2 < 0$, $P < 0.05$). For high SFP scenarios the probability of choosing management regimes generating high dwarf diversity increases ($\beta_3^2 > 0$, $P < 0.05$).

Relative parameter stability to choice set specification shows that this aspect of model estimation has little bearing on the final formulation (Fig. 2). We illustrate this for two sets of relative parameters: (i) the parameter capturing the preference for heather diversity on heather dominated land (β_2^1) relative to the parameter capturing the farmers' propensity to avoid burning costs (β_1^2); (ii) the parameter capturing the change in preference for stocking density under a reduction in the single farm payment to less than £20/ha (β_3^1) relative to the parameter capturing the disutility from occurring away wintering costs (β_1^3).

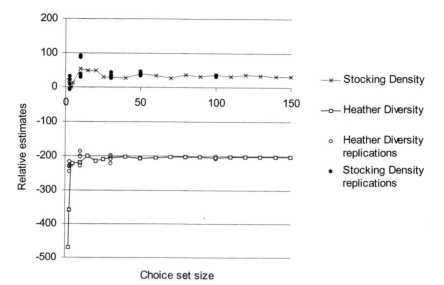

Fig. 2 Relative parameter estimates in choice model related to stocking density β_3^1/β_1^3 and related to heather density β_2^1/β_1^2 as a function of choice set size. Replication of estimations shown for choice set size 3, 10, 30, 50, 100, five replications for each choice set size.

The analysis shows that when the size of the choice set is below approximately 50 the relative parameter values are not stable, but results seem to be unaffected by the size of the choice set for larger choice sets. The results reported (Table 3) and used for further analysis are the results for a model estimated with 100 alternatives.

7. Predicted impact on stocking densities

The generalised linear model fitted to simulated summer stocking densities ($R_{adj}^2 = 0.816$) showed significant effects of the size of the area of the management unit ($t = 2.008$, $P = 0.045$), current mean temperature ($t = 72.30$, $P < 0.001$), degree of warming ($t = 218.0$, $P < 0.001$) and single farm payment ($P < 0.001$ for both factor levels) (Table 4). Farmers apply on average a higher summer stocking density to larger management units with higher temperatures (i.e. areas at lower elevation or with more severe climate change). Low levels of single farm payment are predicted to lead to a decrease in stocking density compared to the current levels (Fig. 3A). High levels of the payment is also predicted to lead to reduced stocking densities

Table 4 Determinants of summer and winter stocking densities in land management units. Summer stocking density model, $R^2_{adj}=0.8161$, $P<0.001$. Winter stocking density model, $R^2_{adj}=0.5161$, $P<0.001$.

Variable	Summer stocking densities		Winter stocking densities	
	Estimates	t-statistics	Estimates	t-statistics
Intercept	-1.862	-44.396	0.183	17.192
SFP_M	0.439	79.922	0.0573	40.985
SFP_H	0.324	58.961	0.0627	44.837
Warming	0.489	217.958	0.0584	102.301
MU area	8.164×10^{-6}	2.008	-8.132×10^{-7}	-0.787
Temperature	0.398	72.303	0.0518	37.018

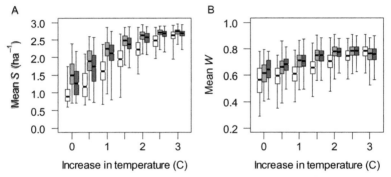

Fig. 3 Responses of grazing strategies to joint variation in the single farm payment (SFP) and climate change. Boxplots show equilibrium (A) summer sheep densities S (ha^{-1}) and (B) winter stocking rates W (white, SFP $<£20$ ha^{-1}; light grey, $£20 <$ SFP $< £60$ ha^{-1}; dark grey, SFP $>£60$ ha^{-1}). Boxplots show medians (thick lines), interquartile ranges (IQRs, boxes), ranges (whiskers) and ±1.58 IQR/$\sqrt{n}$ (notches) where $n=$ number of datapoints of values for the mean values of individual management units. If two notches do not overlap, this indicates that the medians are significantly different at $P<0.05$. Mean values for each management unit are calculated from the strategies chosen during 500 simulated years, following an initial 500-year transient period. Values from 10-replicate simulations are used in the plots.

compared to the current payment levels ($P<0.001$); however, this effect is only apparent when analysing the individual management unit data. The landscape mean over the 10 simulations is not different to the mean outcome under the current payments (Fig. 3A). The simulations suggest that increasing temperature under climate change outweighs the differences resulting from changes in the single farm payment. However, as the management regimes do not include options above a stocking density of 3 sheep/ha, this prediction could be a result of this selection of choice set design (Fig. 3A). Similar results hold for variation in the winter stocking densities; however, the effect of the size of management area is not significant $(t=-0.787, P=0.432)$. Furthermore, the mean effect from the 10 simulations of different levels of the single farm payment also suggests that only the low payment levels generate a significantly lower winter stocking density for some warming scenarios (Fig. 3B).

8. Predicted impact on burning activities

Simulated proportions of time employing grouse moor management (strategies involving some burning) were analysed using a GLM with binomial error structure corrected for overdispersion $(R^2_{adj}=0.405)$ (Table 5). Under the current climate, managers mostly express a clear preference for either grouse moor management or sheep production (Fig. 4A–C). However, a reduced preference for grouse is seen at medium levels of SFP $(t=-31.65, P<0.001)$ and for large management units $(t=-4.083, P<0.001)$ at currently high temperatures/low elevations $(t=-23.41, P<0.001)$. As the climate gets warmer, the preference grouse management also declines $(t=-86.63, P<0.001)$ and we see a shift away from managed burning towards grazing (Fig. 4A–C). However, there is only a small impact on the burning practices of management units that still engage in burning (Fig. 4D).

Table 5 Binomial linear model of participation in managed burning. $R^2_{adj}=0.405$. Correction for over dispersion, $k=78.68075$.

Variable	Estimates	t-statistics
Intercept	4.742	21.780
SFP$_M$	−0.973	−31.619
SFP$_H$	0.0606	2.357
Warming	−1.323	−86.538
MU area	-8.787×10^{-5}	−4.080
Temperature	−0.672	−23.396

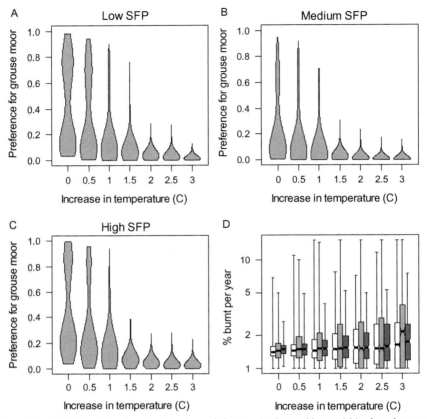

Fig. 4 (A–C) Violin plots showing the kernel density (indicated by width) of preference for grouse moor management for different temperature rises and levels of the single farm payment (SFP). Preference is measured as the proportion of years in which management units employ a strategy involving burning. (D) Boxplot of the mean percentage of the dwarf shrub burnt each year when management units employ strategies including burning, for variations in the single farm payment and climate change (all as in Fig. 3).

9. Predicted impact on land cover

The results on vegetation cover reflect both the natural habitat dynamics and the management responses. At current temperatures the proportions of dwarf shrub are inversely related to the summer grazing pressure determined by the SFP level, while the opposite is true for graminoids and bare peat, both of which are favoured by higher grazing (Fig. 5A–C). Bracken is not eaten by sheep so is relatively unaffected by grazing

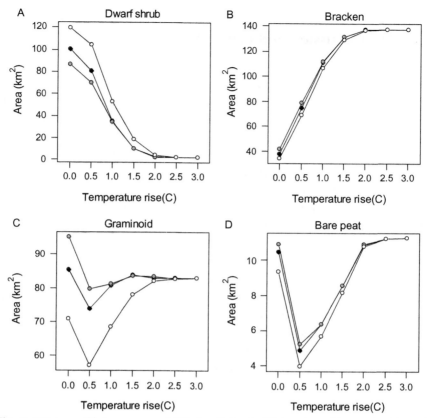

Fig. 5 Mean equilibrium areas of (A) dwarf shrub, (B) bracken, (C) graminoids and (D) bare peat for differing temperature rise and single farm payments (white = low, grey = medium, black = high). Means are values from 10-replicate simulations after 1000 years.

(Fig. 5D). However its habitat quality is strongly constrained by a need for high temperature and so warmer temperatures allow it to move up the hill and increase in cover. Dwarf shrub quality peaks at moderate temperatures, so in theory warming should increase cover in the coldest sites but decrease it in the warmest. However, warmer temperatures also bring higher grazing pressure in the model, which offsets any gains in cover in the colder sites. Graminoids and bare peat both favour colder temperatures, so a slight warming reduces their cover. However, as the temperature increases further and dwarf shrubs are lost to overgrazing then graminoids and bare peat expand in cover. At warmer temperatures, management converges away from burning for grouse moor and towards sheep production and so the responses of vegetation to variation in the SFP become smaller (Fig. 5).

10. Discussion and conclusion

Much work has been devoted to anticipate the affects of alternative reforms of the European agricultural policy schemes. Previous modelling studies from the UK uplands have shown a general trend towards extensification in the uplands as a result of decoupling production and subsidy payments with the introduction of the single farm payment (Matthews et al., 2006; Oglethorpe, 2005; Szvetlana et al., 2008). Extensification includes a reduction in sheep and cattle densities and also a reduction in the average per hectare use of inorganic fertilisers, the movement to lower stocking densities and shifts away from traditional suckler beef systems to sheep systems. The substitution of cattle systems with sheep is thought to be due to the lower fixed costs (i.e. operations and contractor costs) associated with sheep systems (Matthews et al., 2006). The results of this study suggest that even when subsidy payments are decoupled from agricultural production, the level of payments is likely to impact the priorities in land management, although this should not be the case if farmers were strictly profit maximising agents. Our analysis suggests that there are two competing explanations for the nature of the impact on land management of changes to the level of the single farm payments. These two competing explanations could broadly be defined as the payment scheme's impact on: (i) the propensity to diversify to make up for lost earnings, and (ii) the economic capacity of land owners to maintain a given level of land management activity. Our analysis shows that a reduction in the single farm payment leads to a reduction in stocking density and an increase in burning activities. This result can be interpreted by combining the insights from the behavioural model and the vegetation dynamics model. The reduction in the single farm payment reduces the stocking density. This mainly results from the proportion of farmers giving up sheep production under this scenario. In turn, the reduced sheep density promotes dwarf shrub regeneration, and this leads to an increase in the utility from choosing a grouse production management strategy. Overall, the results suggest that reducing the single farm payment introduces a shift in land management priorities to a regime less dominated by sheep grazing and more dominated by grouse shooting. As the single farm payment is independent of the amount of production, this suggests that current management practices are not financially viable but maintained through the subsidy. An increase in the single farm payment also results in a shift away from sheep grazing towards grouse moor management. This suggests that under more favourable financial circumstances, expensive but valued management regimes become relatively more attractive and land managers chose to invest more in the generation of habitats for grouse production. Again this suggest that land managers in the upland are motivated not purely by financial objectives but still are impacted by the financial constraints under which they carry out their business operations. Taken together, these results suggest that "other factors" rather than just profit contribute to farmers' utility, and therefore also important in determining land management decisions.

In terms of understanding these other factors, there are at least two key fields that provide conceptual and empirical insights. The first comes from behavioural studies, considering in particular, why some farmers voluntarily sign up to some policies and not others (Aslam et al., 2017; Brotherton, 1991, 1989; Potter and Gasson, 1988; Wilson, 1997; Zandersen et al., 2016). The second field is based in rural sociology and seeks to account for the persistence of "middle-range" family farms that are too big to be a "hobby" farm but too small to be economically viable on their own. In this case, family structure and employment mobility are seen as important factors that drive the decision to stay in farming (Munton and Marsden, 1991). In our study, we have not been able to quantify socio-economic characteristics favouring particular types of behaviours. However, the more qualitative studies from this study area do give evidence of different land management groupings impacting on land management choices (Dougill et al., 2006). External industry factors, along the supply and demand chain, may also be important factors of land management behaviour, e.g., work in Canada has showed that structural factors such as the nature of the food processing industry (Fraser, 2006) and land tenure (Fraser, 2004) all have significant influences on the types of crops a farmer plants and the extent to which they engage in soil conservation practices.

This study has also investigated the extent to which the interaction between the decision-making in various habitats and the vegetation dynamics is sensitive to climate change. The model predicts significant changes to both land management behaviour and vegetation cover. The system is predicted to shift towards a grazing-based system, where stocking densities increase, dwarf cover declines resulting in a reduction in burning activities. It should however be noted, that in the vegetation models, temperature increases that pushed individual cells above the observed maximum temperature were treated as having the same effect as raising temperature to the observed maximum. This was because nonlinear relationships fitted to local data could not be extrapolated to higher temperatures. This means that bracken may in reality continue to increase in dominance at high temperatures rather than reach an asymptote. Furthermore, the temperature dependency in the behavioural model is represented by a linear relationship, this may have resulted in unrealistic predictions of climate effects for lower elevation areas. Management decisions converge at high temperatures but this could be due to the fact that high stocking density options are not present in the choice set. This means that the increase in stocking densities might be even more severe that the simulated results indicate.

The approach illustrated in this paper could potentially be improved to include more elaborated decision models given replications of the types of data we use here to a larger sample size. The relatively recent developments in discrete choice modelling to capture heterogeneity in preferences between individuals or groups of individuals, by application of the mixed logit model (Train, 2003) or the latent class models (Boxall and Adamowicz, 2002), offer interesting extensions of the approach we have proposed here. Furthermore, new developments in choice modelling over the last decade could potentially be used to model complex decision-making processes. This could potentially be achieved using hybrid choice models (Ben-Akiva et al., 2002; Hess and Beharry-Borg, 2012).

The present paper has however made a contribution to the way in which discrete choice models of land use behaviour can be considered in an integrated model. Our approach allows the simulation of choices in response to the environmental consequences of past actions, a factor usually ignored in studies using choice modelling. Furthermore, the behavioural rules are revealed through empirical data rather than dictated through expert opinions, derived from the literature or simply assumed. This allows the research to test the statistical significance of various determinants of choice. Moreover, the behavioural choice model generates probabilities of alternative behaviours that makes it ideally suited for integration with simulation models.

Acknowledgements

The paper has been funded through the Rural Economy and Land Use (RELU) programme, co-sponsored by Defra and SEERAD (project RES-224-25-0088). The authors would like to thank the farmers, landowners and gamekeepers of the Peak District National Park who gave their time to take part in the questionnaires. Furthermore, the paper has benefited from support from the PREAR project (No 652615) granted under the FACCE SURPLUS ERA-NET Cofund.

References

Agarwal, C., Green, G.M., Grove, J.M., Evans, T.P., Schweik, C.M., 2002. A Review and Assessment of Land-Use Change Models: Dynamics of Space, Time and Human Choice. US Department of Agriculture, Forest Service Northeastern Research Station, Newton Square, PA.

An, L., 2012. Modeling human decisions in coupled human and natural systems: review of agent-based models. Ecol. Model. 229, 25–36.

Armstrong, H.M., Gordon, I.J., Grant, S.A., Hutchings, N.J., Milne, J.A., Sibbald, A.R., 1997a. A model of the grazing of hill vegetation by the sheep in the UK. I. The prediction of vegetation biomass. J. Appl. Ecol. 34, 166–185.

Armstrong, H.M., Gordon, I.J., Hutchings, N.J., Illuis, A.W., Milne, J.A., Sibbald, A.R., 1997b. A model of the grazing of hill vegetation by sheep in the UK. II. The prediction of offtake by sheep. J. Appl. Ecol. 34, 186–207.

Aslam, U., Termansen, M., Fleskens, L., 2017. Investigating farmers' preferences for alternative PES schemes to promote carbon sequestration in UK agroecosystems. Ecosyst. Serv. 27 (A), 103–112.

Bailey, K.D., 1982. Methods of Social Research. Free Press, New York.

Balmann, A., 1997. Farm-based modelling of regional structural change. Eur. Rev. Agric. Econ. 25 (1), 85–108.

Baltas, N.C., Korka, O., 2002. Modelling farmers' land use decisions. Appl. Econ. Lett. 9, 453–457.

Barclay-Estrup, P., Gimingham, C.H., 1969. Description and interpretation of cyclical processes in a heath community. I. Vegetational change in relation to the *Calluna* cycle. J. Ecol. 57, 737–758.

Barreteau, O., Bousquet, F., 2000. SHADOC: a multi-agent model to tackle viability of irrigated systems. Ann. Oper. Res. 94, 139–162.

Ben-Akiva, M., McFadden, D., Train, K., Walker, J., Bhat, C., Bierlaire, M., Bolduc, D., Boersch-Supan, A., Brownstone, D., Bunch, D.S., Daly, A., De Palma, A., Gopinath, D., Karlstrom, A., Munizaga, M.A., 2002. Hybrid choice models: progress and challenges. Mark. Lett. 13 (3), 163–175.

Berger, T., 2001. Agent-based spatial models applied to agriculture: a simulation tool for technology diffusion. Resource use changes and policy analysis. Agric. Econ. 25 (2–3), 245–260.

Bousquet, F., Bakam, I., Proton, H., Le Page, C., 1998. Cormas: commonpool resources and multi agent systems. In: Lecture Notes in Artificial Intelligence. 1416, Springer, pp. 826–837.

Boxall, P., Adamowicz, W., 2002. Understanding heterogeneous preferences in random utility models: a latent class approach. Environ. Resour. Econ. 23, 421–446.

Brock, W.A., Durlauf, S.N., 2006. Multinomial choice with social interactions. In: Blume, L.E., Durlauf, S.N. (Eds.), The Economy as an Evolving Complex System, III. Current Perspectives and Future Directions. Santa fee Institute Studies in the Sciences of Complexity, pp. 175–206.

Brotherton, I., 1989. Farmer participation in voluntary land diversion schemes: some observations from theory. J. Rural. Stud. 5, 299–304.

Brotherton, I., 1991. What limits participation in ESAs? J. Environ. Manag. 32, 241–249.

Chapman, D., Termansen, M., Quinn, C.H., Jin, N., Bonn, A., Cornell, S., Fraser, E.D.G., Hubacek, K., Kunin, W., Reed, M., 2009. Modelling the coupled dynamics of moorland management and vegetation in the UK uplands. J. Appl. Ecol. 46, 278–288.

Chapman, S.S., Bonn, A., Kunin, W.E., Cornell, S.J., 2010. Random forest characterization of upland vegetation and management burning from aerial imagery. J. Biogeogr. 37, 37–46.

COM, 2017. The Future of Food and Farming. COM, Brussels. 29.11.2017. (2017) 713 final.

Conrad, S.A., Yates, D., 2018. Coupling stated preferences with a hydrological water resource model to inform water policies for residential areas in the Okanagan Basin, Canada. J. Hydrol. 564, 846–858.

Cooper, J.C., Keim, R.W., 1996. Incentive payments to encourage farmer adoption of water quality protection practices. Am. J. Agric. Econ. 78 (1), 54–64.

Dia, H., 2001. An agent-based approach to modelling driver route choice behaviour under the influence of real-time information. Transp. Res. C 10, 331–349.

Dougill, A.J., Fraser, E.D.G., Holden, J., Hubacek, K., Prell, C., Reed, M.S., Stagl, S., Stringer, L.C., 2006. Learning from doing participatory rural research: lessons from the Peak District National Park. J. Agric. Econ. 57, 259–275.

Drechsler, M., Johst, K., Ohl, C., Wätzold, W., 2007. Designing cost-effective payments for conservation measures to generate spatiotemporal habitat heterogeneity. Conserv. Biol. 21, 1475–1486.

English Nature, 2003. England's Best Wildlife and Geological Sites; The Condition of SSSIs in England in 2003. English Nature, Peterborough.

Evans, T.P., Kelley, H., 2004. Multi-scale analysis of a household level agent-based model of land cover change. J. Environ. Manag. 72, 57–72.

Fraser, E.D.G., 2004. Land tenure and agricultural management: soil conservation on rented and owned fields in southwest British Columbia. Agric. Hum. Values 21, 73–79.

Fraser, E.D., 2006. Crop diversification and trade liberalization: linking global trade and local management through a regional case study. Agric. Hum. Values 23, 271–281.

Hess, S., Beharry-Borg, N., 2012. Accounting for latent attitudes in willingness-to-pay studies: the case of coastal water quality improvements in Tobago. Environ. Resour. Econ. 52, 109–131.

Huber, R., Bakker, M., Balmann, A., Berger, T., Bithell, M., Brown, C., Gret-Regamey, A., Xiong, H., Le, Q.B., Mack, G., Meyfroidt, P., Millington, J., Muller, B., Polhill, J.G., Sun, Z., Seidl, R., Troost, C., Finger, R., 2018. Representation of decision-making in European agricultural agent-based models. Agric. Syst. 167, 143–160.

Huigen, M.G.A., 2004. First principles of the Mameluke multi-actor modelling framework for land use change, illustrated with a Philippine case study. J. Environ. Manag. 72, 5–21.

Hunt, L.M., Kushneriuk, R., Lester, N., 2007. Linking agent-based and choice models to study outdoor recreation behaviours: a case of the landscape fisheries model in northern Ontario, Canada. For. Snow Landsc. Res. 81, 163–174.

Lynch, L., Hardie, I., Parker, D., 2002. Analyzing Agricultural Landowners'Willingness to Install Streamside Buffers. Working Paper from, The Department of Agricultural and Resource Economics, The University of Maryland, College Park.

Matthews, K.B., Wright, I.A., Buchan, K., Davies, D.A., Schwarz, G., 2006. Assessing the options for upland livestock systems under CAP reform: developing and applying a livestock systems model within whole-farm systems analysis. Agric. Syst. 90, 32–61.

Milon, J.W., Scrogin, D., 2006. Latent preferences and valuation of wetland ecosystem restoration. Ecol. Econ. 56, 162–175.

Munton, R., Marsden, T., 1991. Dualism or diversity in family farming? Patterns of occupancy change in British agriculture. Geoforum 22, 105–117.

Nainggolan, D., Hasler, B., Andersen, H.E., Gyldenkærne, S., Termansen, M., 2018. Water quality management and climate change mitigation: cost-effectiveness of joint implementation in the Baltic Sea region. Ecol. Econ. 144, 12–26.

Office for National Statistics, 2003. Census 2001: CD Supplement to the National Report for England and Wales and Key Statistics for Local Authorities in England and Wales. Office for National Statistics, London.

Oglethorpe, D.R., 2005. Livestock production post CAP reform: implications for the environment. Anim. Sci. 81, 189–192.

Palmer, S.C.F., Gordon, I.J., Hester, A.J., Pakeman, R.J., 2004. Introducing spatial grazing impacts into the prediction of moorland vegetation dynamics. Landsc. Ecol. 19, 817–827.

Parsons, G.R., Kealy, M.J., 1992. Randomly drawn opportunity sets in a random utility model of lake recreation. Land Econ. 68, 93–106.

Peak District National Park, 2004. State of the Park Report (Update). http://www.peakdistrict.org. Peak District National Park Authority.

Peterson, J.M., Fox, J.A., Leatherman, J.C., Smith, C.M., 2007. Choice experiments to assess farmers' willingness to participate in a water quality trading market. In: Selected Paper Prepared for Presentation at the American Agricultural Economics Association Annual Meeting. Kansas State University, Portland, Oregon. July 29-August 1.

Polhill, J.G., Gotts, N.M., Law, A.N.R., 2001. Imitative versus nonimitative strategies in a land-use simulation. Cybern. Syst. 32 (1), 285–307.

Potter, C., Gasson, R., 1988. Farmer participation in voluntary land diversion schemes: some predictions from a survey. J. Rural. Stud. 4, 365–375.

Prestele, R., Arneth, A., Bondeau, A., de Noblet-Ducoudré, N., Pugh, T.A.M., Sitch, S., Stehfest, E., Verburg, P.H., 2017. Current challenges of implementing anthropogenic land-use and land-cover change in models contributing to climate change assessments. Earth Syst. Dynam. 8, 369–386.

Schreinemachers, P., Berger, T., 2011. An agent-based simulation model of human-environment interactions in agricultural systems. Environ. Model. Softw. 26 (7), 845–859.

Shaikh, S.L., Sun, L., van Kooten, G.C., 2005. Are agricultural values a reliable guide in determining Landowners'decisions to create forest carbon sinks? Can. J. Agric. Econ. 55, 97–114.

Shannon, C.E., 1948. A mathematical theory of communication. Bell Syst. Tech. J. 27, 379–423 and 623–656.

Sustainable Uplands and Moors for the Future, 2007. Looking After Moorland Habitats. Sustainable Uplands Project. Research Note 14. School of Earth and Environment, University of Leeds, UK.

Szvetlana, A., Hanley, N., Dallimer, M., Robertson, P., Gaston, K., Armsworth, P.R., 2008. Impacts of policy reform on sustainability of hill farming in U.K. by means of bio-economic modelling. In: Paper Prepared for Presentation at the 107th EAAE Seminar "Modelling of Agricultural and Rural Development Policies". Sevilla, Spain.

Takama, T., Preston, J., 2008. Forecasting the effects of road user charge by stochastic agent-based modelling. Transp. Res. A 42, 738–749.

Termansen, M., McClean, C.J., Skov-Petersen, H., 2004. Recreational site choice modeling using high spatial resolution data. Environ. Plan. A 36, 1085–1099.

Tieskens, K.F., Shaw, B.J., Haer, T., Schulp, C.J.E., Verburg, P.H., 2017. Cultural landscapes of the future: using agent-based modeling to discuss and develop the use and management of the cultural landscape of South West Devon. Landsc. Ecol. 32, 2113–2132.

Train, K., 2003. Discrete Choice Methods With Simulation. Cambridge University Press.

Train, K., 2009. Discrete Choice Methods With Simulation, Second ed. Cambridge University Press.

Train, K., 2016. Mixed logit with a flexible mixing distribution. J. Choice Model. 19, 40–53.

UK Biodiversity Steering Group, 1995. The UK Steering Group Report—Volume II: Action Plans. HMSO, London.

Vanslembrouck, I., Huylenbroeck, G.V., Verbeke, W., 2002. Determinants of the willingness of Belgian farmers to participate in agri-environmental measures. J. Agric. Econ. 53 (3), 489–511.

Vedel, S.E., Jacobsen, J.B., Thorsen, B.J., 2015. Contracts for afforestation and the role of monitoring for landowners' willingness to accept. Forest Policy Econ. 51, 29–37.

Veldkamp, A., Verburg, P.H., 2004. Modelling land use change and environmental impact. J. Environ. Manag. 72, 1–3.

Watson, A., Miller, G.R., 1976. Grouse Management. Game and Wildlife Conservation Trust, Hampshire: Fordingbridge.

Wilson, G.A., 1997. Factors influencing farmer participation in the environmentally sensitive areas scheme. J. Environ. Manag. 50, 67–93.

Wolff, S., Schrammeijer, E.A., Schulp, C.J.E., Verburg, P.H., 2018. Meeting global land restoration and protection targets: what would the world look like in 2050? Glob. Environ. Chang. 52, 259–272.

Zandersen, M., Jørgensen, S.L., Nainggolan, D., Gyldenkærne, S., Winding, A., Greve, M.H., Termansen, M., 2016. Potential and economic efficiency of using reduced tillage to mitigate climate effects in Danish agriculture. Ecol. Econ. 123, 14–22.

Further reading

Herriges, J.A., Phaneuf, D.J., 2002. Inducing patterns of correlation and substitution in repeated logit models of recreation demand. Am. J. Agric. Econ. 84, 1076–1090.

Silvano, R.A.M., Silva, A.L., Ceron, M., Begossi, A., 2008. Contributions of ethnobiology to the conservation of tropical rivers and streams. Aquat. Conserv. Mar. Freshwat. Ecosyst. 18 (3), 241–260.

Willock, J., Deary, I.J., Mcgregor, M.M., Sutherland, A., Edward-Jones, G., Morgan, O., Dent, B., Grieve, R., Gibson, G., Ausin, E., 1999. Farmers' attitudes, objectives, behaviors, and personality traits: the Edinburgh study of decision making on farms. J. Vocat. Behav. 54, 5–36.

CHAPTER FIVE

Differing perceptions of socio-ecological systems: Insights for future transdisciplinary research

Noa Avriel-Avni[a,*], Jan Dick[b]

[a]Dead Sea and Arava Science Center, Masada, Israel
[b]Centre for Ecology and Hydrology, Penicuik, Scotland, United Kingdom
*Corresponding author: e-mail address: noa@adssc.org

Contents

Abstract

The growing understanding that transdisciplinary research is required for sustainable land management (i.e., co-production of knowledge by researchers and land managers) stems from the complexity and unpredictability of social-ecological systems. However,

Advances in Ecological Research, Volume 60
ISSN 0065-2504
https://doi.org/10.1016/bs.aecr.2019.03.001

many scientists feel that the large gap between the agendas and worldviews of scientists and land managers makes it difficult to co-produce knowledge. This challenge was the focus of our study in Cairngorms National Park (CNP), Long-Term Social-Ecological Research Platform (LTSER), Scotland.

Semi-structured interviews were conducted with 18 land managers and 15 scientists, who are active in CNP, focussed on their individual perception of the park's social-ecological system. The findings point to differences in interests between the two groups. Land managers are mainly troubled by local economic and legacy problems, while scientists are more concerned by environmental and global questions. However, the findings also indicated a shared sense of uncertainty about the future of the region along with willingness for both groups to work together. These findings suggest a need for transdisciplinary research that co-produces science best future vision; i.e., a synthesis of scientific knowledge and land managers' practical knowledge, motivations and aspirations to create a resilient socio-ecological system.

1. Introduction

Decision-making in land management has a major impact on the environment as recognized by major global initiatives such as the Millennium Ecosystem Assessment (MEA, 2005), The Economics of Ecosystems & Biodiversity (TEEB, 2010), and the more recent Intergovernmental Platform on Biodiversity and Ecosystem Service (IPBES) (Díaz et al., 2015). Reduction of carbon sequestration due to deforestation (Deary and Warren, 2017), amplification of nitrogen emissions and the contamination of water sources due to improper handling of animal effluent (Fowler et al., 2013) as well as the loss of local biodiversity (Carmen et al., 2018) are well-studied examples of the negative impacts of inappropriate land management. On the other hand, agriculture itself is a risk-management business, often of limited resilience, as was highlighted in a recent report of the Scottish Parliament (Thomas, 2018). One way to deal with such threats and risks is the EU stipulation of agricultural subsidies in compliance with stringent environmental standards (Sutherland, 2010). As effective as this approach has been, however, subsides have excluded land managers from the process of decision-making and have not taken into account the complexity of the social-ecological system in which they make decisions (Bai et al., 2016; Canova et al., 2019; Teschner et al., 2017; Wamsler, 2017).

Socio-ecological systems (SESs) are inherently complex systems (Dearing et al., 2010; Folke et al., 2010; Lambin et al., 2001). This complexity is enhanced by globalization, which leads to relatively free flow of information, knowledge, materials, and people between local SES across global scales (Ericksen, 2008). Globalization exacerbates the disconnection between areas where resources are harvested (suffering environmental impact) and those where they are consumed (enjoying the production). Thus, globalization may weaken the responsibilities of land managers for ecosystems that provide ecosystem services (Canova et al., 2019). These processes have been accelerating (Plummer and Armitage, 2007), resulting in SES that are extremely open and much more complex (Costanza et al., 2014). Consequently, SES are changing rapidly in terms of land use (e.g., a transition from sheep breeding farms to multiple land uses), social and economic structures (e.g., transition to international regulation) and human activity, such as tourism in places that were once only agricultural lands (Ruhl et al., 2007). These processes of change increase the uncertainties of SES management (Costanza et al., 1993; Haasnoot et al., 2013; Holden et al., 2014) and raise urgent questions about sustainability and resilience at local (Teschner et al., 2017) and global scales (Costanza et al., 2014; Folke et al., 2010; Haasnoot et al., 2013; Heck et al., 2018; Holden et al., 2014). The consequences of climate change, which are inherently difficult to predict, increase this uncertainty still further (Costanza et al., 2014; Ostrom, 2010; Wise et al., 2014).

The concept of sustainability has many interpretations and has been widely criticized in recent years. Holden et al. (2014) showed that the concept remains relevant, however, and that it can be evaluated in social-environmental systems. In this article we adopt the SES model, which postulates that social systems and life-supporting ecological systems are interdependent (Folke et al., 2010). Sustainable development, within the framework of the SES model, is not necessarily the preservation of the existing system. Instead, it describes the ability of the SES to tolerate unknown or unforeseen shocks via absorbing, accommodating, or implementation changes (adaptation) to the impact, or by fundamentally reorganizing the SES in response to the challenge in ways that were impossible within the existing SES state (transformation) (Barnes et al., 2017; Folke et al., 2010). The continuity of the social-ecological system can therefore be described as a process of transformation or adaptation to the external change. However, the social part of the SES has additional learning, design, and management capabilities that can markedly shorten the adaptation period and increase the resilience of the coupled systems (Plummer and Armitage, 2007). A resilient system is one that recovers quickly to its pre-disrupted condition or is reorganized into a new stable state (MEA, 2005; see chapter "Adaptive capacity in ecosystems" by Angeler et al.). Thus, sustainability is a dynamic process not a fixed, static state (Turner et al., 2008). Sustainability is dependent, according to the SES model, on human behaviour that reduces the pressure on life-supporting ecological systems and at the same time controls the exploitation of ecosystem services (Folke et al., 2010). The interdependence of the coupled elements of the system is the reason for the pronounced impact of decisions taken by stakeholders about the management and development of their local system (Bai et al., 2016). In the face of this complex reality and the great uncertainty regarding the progression of local and global processes, however, stakeholders have objective difficulties in planning and maintaining SES sustainability (Ericksen, 2008; Haasnoot et al., 2013; Ostrom, 2010).

To address this challenge, scientists have recently been called upon to harness their research skills to mitigate environmental problems (Janssen et al., 2006; Magliocca et al., 2018; Ostrom, 2010). Within the scientific community there is an understanding that sustainable and resilient solutions need to be formulated by collaboration between scientists and stakeholders (Glass et al., 2013; Mauser et al., 2013; Mielke et al., 2016). Such research requires scientists to use transdisciplinary approaches (Angelstam et al., 2018; Dick et al., 2018a; Holzer et al., 2018; Mirtl et al., 2018; Plummer et al., 2017; van der Hel, 2016) to create "problem-focussed groups" composed of researchers from different disciplines and those relevant stakeholders to co-design and co-deliver theoretical and practical knowledge that solves socio-ecological problems (Avriel-Avni et al., 2017; Dick et al., 2017; Glass et al., 2013; Holzer et al., 2018; Jax et al., 2018; Moore, 2013; van der Hel, 2016).

Scientists who practice transdisciplinary approaches face other challenges, however, such as the frustration that a substantial gap exists between their scientific understanding of environmental management and that of other stakeholders, which makes it difficult to form a fruitful collaboration (Teschner et al., 2017; Thompson et al., 2017). This challenge to building knowledge for sustainable management in the process of transdisciplinary research was the focus of our study. The Cairngorms National Park (CNP) was chosen as a case study because it is a LTSER platform (see below) and because the Management Board of the park had interest in cooperating with land managers in leading sustainable management (Evely et al., 2008).

1.1 From problem-based research to international research networks

Urgent environmental challenges require scientific researchers to carry out problem-based research (Janssen et al., 2006; Jax et al., 2018; Ostrom, 2010) and actionable science (Mauser et al., 2013) in order to develop models of sustainability and system resilience (Magliocca et al., 2018; van der Hel, 2016; Wise et al., 2014). In order to transform local solutions into principles of sustainable management, generalizations need to be made. The formulation of global generalizations should be based on multiple local case studies (Barnes et al., 2017; Ostrom, 2010) and demands the creation of research networks based on focal points of change (Barnes et al., 2017; Lambin et al., 2001). Although this approach is logical, creating generalizations that rely on case studies requires careful methodical work (Magliocca et al., 2018). Several research initiatives across the globe, such as the "Future Earth" program (Mauser et al., 2013; van der Hel, 2016) and time-limited research projects such as the OpenNESS project (Jax et al., 2018), have adopted this approach (Barnes et al., 2017).

1.2 The network of long-term social-ecological research (LTSER)

The LTSER platform network is one of the leading, long-term examples of this type of transdisciplinary collaboration. The network consists of place-based socio-ecological research groups. In Europe, these groups have been built around key SES. Historically, this network is an evolution of the global network of Long-Term Ecological Research stations (LTER). The transition from LTER to LTSER network was accompanied by a transformation from a focus on conceptualization toward implementation research (Mirtl et al., 2018). Collaboration between LTSER platforms is expressed in the monitoring of biophysical and social parameters through joint protocols, sharing knowledge between research platforms as well as creating common generalizations about social-ecological issues (Dick et al., 2018b; Holzer et al., 2018; Mirtl et al., 2018).

1.3 Design of socio-ecological solutions together with stakeholders: Transdisciplinarity

Given the open nature of the SESs, land managers often encounter difficulty making decisions that will increase the resilience of their local SES. Food production, which is a major land use in the CNP, can be used to demonstrate this challenge. Food production, controlled by land managers, is only one part of the long food supply chain that links food producers to consumers (Ericksen, 2008). Given this complexity, land managers find it difficult to make knowledge-based decisions when faced with external shocks, but their subsequent behaviour and decisions directly influence the local and global environment (Bai et al., 2016). In order to cope with this complexity, there is a growing recognition of the need to build transdisciplinary research frameworks, involving stakeholders and interdisciplinary teams of natural and social scientists (Barlow et al., 2011; Haughton et al., 2009), even though interdisciplinary working can present many challenges to classical scientific approaches (Evely et al., 2008; Janssen et al., 2006; van der Hel, 2018). In light of the understanding that groups of stakeholders within a social system play a vital role in the SES structure and function (Folke, 2006; Turner et al., 2008), there are multiple voices calling for cooperation between researchers and stakeholders in order of design adaptive and sustainable systems (Barnes et al., 2017; Belmont Forum, 2011; Jax et al., 2018; Krasny and Roth, 2010; Löf, 2010; Popa et al., 2015; van der Hel, 2016).

Inspired by the concept of SES as dynamic systems, the European LTER network adopted a transdisciplinary approach through the establishment of Long-Term Socio-Ecological Research (LTSER) platforms. The aim was to engage scientists from multiple disciplines working together (interdisciplinary) with multiple local stakeholders in the process of understanding the socio-ecological system (Dick et al., 2018b; Haberl et al., 2006; Holzer et al., 2018; Mirtl et al., 2018). In Scotland, for example, earlier studies called for greater collaboration between the actors involved in land management and regulation of the environment in order to ensure sustainable land management (Evely et al., 2008; Glass et al., 2013).

Collaboration between stakeholders and scientists can be at the level of monitoring and assessment of the situation (learning the system) and designing solutions (system adaptation) in order to achieve sustainability (Glass et al., 2013; Haasnoot et al., 2013). The necessity of collaboration becomes clear in light of accumulated experience that suggests that outside intervention, even when it is based on robust scientific knowledge, has generally failed to generate sustainable solutions (Zuber-Skerritt, 2012). This may be due to a failure of scientists (or other intervening entities) to understand the social dynamics that preserve the existing undesirable situation (Bruns and Worsley, 2015) or a failure to harness the internal forces in the social system to bring about the desired change (Bai et al., 2016; Burns, 2007). Moreover, creating practical knowledge demands a process in which actors reflect upon underlying values, norms, or frames and alter their behaviour accordingly (Argyris and Schön, 1974).

1.4 Transdisciplinary research for building a future-vision

Changing how people operate and manage their SES also involves creating an image for their desired future (Bohnet and Smith, 2007; Hermans et al., 2007; Scott, 2011). Developing a "Future-vision," a methodology usually used in psychotherapy and in organizational consulting, can encourage people to start to work for that future. The need to attach some broader significance to one's life is a basic human drive that is particularly important in a rapidly changing world (Levin, 2000). However, the future-vision that incorporates such broad meaning demands a research approach that is scientifically challenging (Bruns and Worsley, 2015; van der Hel, 2016) and this may be why scientific collaboration with stakeholders is still relatively rare (Angelstam et al., 2018; Barnes et al., 2017; Jax et al., 2018; Mielke et al., 2016).

2. Objectives and methodology

2.1 The overall objective

The overall goal of this research in the LTSER was to find a common basis among land managers and scientists for transdisciplinary research. It is based on the assumption that sustainable land management can be achieved through a combination of practical and scientific knowledge. We adopted the theoretical framework of social-ecological systems (SESs), which describes the social and the ecological systems as coupled by mutual influences and eliminates the difference between internal change forces and external drivers (Folke et al., 2010). A comparison of the perceptions of land managers and scientists can enable identification of the interface between the two groups and the possibilities for cooperation between them.

2.2 Study site

Cairngorms National Park (CNP) is in the Scottish Highlands. It is the largest national park in the United Kingdom, measuring 4528 km^2.

The Cairngorms National Park covers parts of the five administrative areas of Aberdeenshire, Moray, Highland, Angus and Perth and Kinross. Over 18,000 people live and work in the park and around 1.7 million people visit the park every year from all over the world (Cairngorms National Park Partnership Plan 2012–2017). The majority of the land is owned by private landowners (including conservation bodies), and people continue to live and work in the park. The historic land-ownership structure in Scotland adds complexity to the socio-ecological system in CNP with large areas owned by an individual, but managed by tenants (Glass et al., 2013). In addition, nearly half of the land in the CNP is considered "wild land" and has been recognized as having international importance for nature and is protected by European Law. Within the CNP there are 19 Areas of Conservation, 12 Special-Protection Areas and 46 Sites of Special-Scientific Interest.

Scotland's National Parks share four aims set out in the National Parks (Scotland) Act 2000, which are: (i) To conserve and enhance the natural and cultural heritage of the area; (ii) To promote sustainable use of the natural resources of the area; (iii) To promote understanding and enjoyment (including enjoyment in the form of recreation) of the special qualities of the area by the public; and, (iv) To promote sustainable economic and social development of the area's communities.

2.2.1 The Cairngorms National Park LTSER platform

The CNP was established in 2003 and extended to its current size in 2010. 2010 was also the year in which the proposal to create an LTSER platform in the Cairngorms National Park was first proposed. It was not until 2013, however, that the five founding institutions signed a Memorandum of Understanding to work together formally. Of the five founding institutions, three were research focussed (Centre for Ecology & Hydrology, the James Hutton Institute, Highlands and Islands University), one was a business (the Crown Estate, who also represented the Scottish Land and Estate sector) and the fifth was the Cairngorms National Park Authority (CNPA, who are responsible for the management of the park). All parties recognized the advantages of a formal institution and the timing was advantageous as the CNPA wanted to establish a research strategy for the park focussed on practical issues, which built on previous research. The boundary of the LTSER platform was agreed to exactly match the park boundary and it encompassed four of the previous LTER sites.

2.2.2 Major environmental issues in Cairngorms National Park

The CNPA, in collaboration with the LTSER researchers, published a Research Strategy (Cairngorms National Park Research Strategy, 2015) composed of four strands: (i) To inform the management of the National Park by connecting research with management needs and providing data to monitor the State of the Park including long-term trends, changes and risks; (ii) To connect research across disciplines and encourage place-based integration; (iii) To facilitate effective knowledge exchange connecting research and practice; and (iv) To promote the Cairngorms National Park as a significant focus for collaborative research contributing to national and international research agendas. The strategy focussed on 20 research questions that each aimed to provide practical knowledge to manage the park. In accordance with the Europe LTSER network approach, one of the main goals was to conduct transdisciplinary research on the CNP LTSER platform (Cairngorms National Park Research Strategy, 2015, p. 2). The study presented in this paper took place in November 2017, prior to the withdrawal from the European Union (Brexit) and this temporal setting of the study may have influenced the responses of the interviewees Fig. 1.

Fig. 1 The light grey area near the top of the map indicates the location of the Cairngorms National Park Authority in United Kingdom. *Source: CNPA site:* http://cairn gorms.co.uk/.

2.3 Data collection

Data collection was accomplished through semi-structured interviews. Two tools were used to encourage the interviewees to share their perceptions and opinions about the local SES with us:

1. A conceptual map of the actors—A "half-baked" map that showed the main actors in the social-ecological system was presented. This tool was designed to reveal how respondents perceive the local SES and their role within it. The interviewees were asked to identify their role on the map, to add more actors and to draw and explain the connections between themselves and relevant actors (see example in Supplemental Materials available online at https://doi.org/10.1016/bs.aecr.2019.03.001).

2. A Likert attitudes questionnaire—The questionnaire contained three blocks of statements about the obstacles, opportunities, type of knowledge required for sustainable land management and importance of collaboration between scientists and land managers. Each interviewee was asked to mark his or her opinion on each statement, from strongly agree to strongly disagree (on a four-point scale). The interviewee also had the opportunity to record that they "had no opinion." We also asked the interviewees to add comments on their choices in a "free-text" box. For the land managers, the questionnaire included specific questions about their views on the situation within the SES. These questions were reformulated for scientists as expressions to reflect their understanding of what land managers think. For example, a statement in the block that dealt with the willingness of land managers to cooperate with scientists was formulated as: "I am willing to cooperate with scientists only if I am assured that I will not lose money." For scientists, this statement was rephrased as: "Farmers would like to cooperate with scientists only if they are assured that they will not lose money." All scientists were asked three additional questions focussed on their views of transdisciplinary research: (1) I believe that farmer knowledge is vital to sustain farming in Scotland; (2) Co-produced solution involving scientist and farmers is vital to sustain farming in Scotland; and (3) Most of my research is co-production of knowledge with farmers. The complete questionnaires appear in the supplementary materials. During the introductions at the start of the interview the aims, method and deliverable were shared verbally with the participants. Formal signing of consent forms was explicitly not done because it was considered culturally insensitive and counter to the claim that participation in the survey was anonymous.

2.3.1 Sample population

Our sample included 18 land managers and 15 scientists. Sim et al. (2018) stated that there is no consensus as to optimum sample size in qualitative research. The literature reports sample sizes of between 5 and 35, depending on many factors including methodological considerations, such as the nature and purpose of the individual study and the epistemological stance underpinning it, but also practical considerations around time and resources. Our sample size was based on the principle of saturation, namely, that new interviewees do not add more information.

1. Land managers were randomly selected from the Yellow Pages telephone directory or were approached on the recommendation of interviewees (snowball approach, Noy, 2008). We adopted this method to avoid bias in selecting interviewees according to the researchers' prior familiarity with them and to ensure a random selection of interviewees. The interviews were usually held in the farmer's homes and lasted between half an hour and an hour, although some lasted up to 2h. In cases where both spouses were involved in the farm, we interviewed both such that they completed the questionnaire separately and then had a joint conversation about the individual questions and the actors map. See Table 1 for details of interviewees.

2. The scientists were from five different research institutions: the Centre for Ecology and Hydrology, the James Hutton Institute, Scotland's Rural College, the Highlands and Islands University and the Moredon Research Institute. They were selected according to their fields of research and the connection with land managers in the CNP, and priority was given to scientists who work in the CNP (see Table 2 for details). The interviews usually lasted about half an hour. Two of the interviews were carried out by telephone

2.4 Data analysis

The interviews were recorded to enable us to re-examine the comments and insights that were expressed during the interviews. Content analysis (Vaismoradi et al., 2016) of the interviews and notes in the researcher's logbook was undertaken in order to identify key themes in the land managers' and scientists' discourse. An emphasis was placed on how the interviewees conceptualize the local SES and how they perceive the possibility of cooperating to meet challenges in the sustainable management of this system.

Table 1 Data on the land managers (LM) interviewed gender (F = female and M = male).

Land manager (LM)	Gender	Age bracket	Farmland	Comments
1	M	60–70	Owner	Also represents farmers in a national union
2	M	60–70	Dweller	Sold their land upon retirement—retained as consultant
3	F	60–70	Dweller	Wife of LM2
4	M	60–70	Owner	The owner of a large estate with a mixed farm. Part of the land is leased to other people
5	M	50–60	Owner	Large estate owner
6	F	70–80	Owner	Mixed farm. They hired someone to manage the land, because of their age
7	M	70–80	Owner	Husband of LM6
8	M	60–70	Farm manager	Mixed sheep and cattle farm
9	M	70–80	Croft-tenant	Mixed farm, including pig farming
10	F	70–80	Croft-tenant	Wife of tenant farmer—who is also very active on the farm
11	F	50–60	Tenant	Member and represents residents on CNP Steering Committee. Active farmers wife
12	M	70–80	Owner	Sheep farmer
13	F	60–70	Owner	Wife of LM12
14	F	15–18	Dweller	Granddaughter of LM12–13
15	M	15–18	Dweller	Grandson of LM12–13
16	M	30–40	Owner	Son of LM12–13
17	M	30–40	Tenant and owner	Sheep farmer
18	M	30–40	Tenant	Part time sheep farmer. Works in CNP Authority

Table 2 Research area of scientists (S) interviewed from the five research institutions.

S#	Research area
1.	Livestock production and knowledge delivery
2.	Poultry specialist
3.	Nitrogen based air pollution from agriculture
4.	Catchment management. LTER manager. EU policy and farmers
5.	Modelling socio-ecological systems
6.	Sociology of agriculture
7.	Applied biodiversity and natural resource management
8.	Insect ecology and ecosystem services
9.	Spatial analysis and species conservation and ecosystem services
10.	Land use and ecosystem service scenarios
11.	Conservation management/sustainable land use
12.	Agricultural ecology
13.	Agricultural economist
14.	Landscape ecologist
15.	Impacts of catchment management on water quality

The analysis proceeded in four stages via repeated reading of the interview texts: (1) Initialization—focussing on the most important constructs which are presented in data and the main issues in the phenomenon under study; (2) Construction—organizing the codes in labelled and more abstract clusters; (3) Rectification—verification of the developed themes by critical examination of the themes; and (4) Finalization—developing the "storyline" or the narrative regarding the SES management, barriers and opportunities for sustainable management. A descriptive statistical analysis was conducted for the questionnaires averaging responses per question separately for the land managers and the scientists.

3. Results

3.1 The SES of Cairngorm National Park

All land managers and scientists interviewed agreed that the Cairngorm National Park is a complex and multi-player, social-ecological system, but these groups expressed this in different ways. The "conceptual map of the actors" created by the interviewees indicated that there are many actors that influence the functioning of the system at local, national and global scales. Land managers predominantly mentioned national and global economic organizations, of politicians, consumers, supermarkets and regulatory agencies, as the external factors affecting their operations. The UK Government was noted in this context because the adoption of the European open market policy exposes farmers to competition with cheaper meat importers from outside the United Kingdom which do not need to comply with EU regulations. Such policies do not protect land managers despite introducing subsidies that encourage local agriculture as part of a food independence policy, for example. Regulation was also mentioned with respect to European environmental regulations that are not well adapted to the local climatic conditions in Scotland.

Supermarkets were described as bodies that dictate the timing of produce sales and drive Scottish farmers to lower prices to meet those of overseas competitors. Government officials enforcing compliance with agricultural subsidy rules were usually viewed as problematic actors due to the burden of administration and work needed to comply, particularly for the protection of biodiversity. Agricultural consultants, on the other hand, were mentioned as key actors in providing appropriate responses to tangible problems. Within the inner circle of the map, the major landowner or Laird, was identified as a key actor in the system. The younger generation was mentioned as a group of actors undermining the future of the farm, since many of them are not interested in managing the land. Due to discontinuous functioning or economic considerations, some farms have been sold to urban residents, either for use as vacation homes or as domiciles. This trend of external ownership was noted to harm the social-agricultural fabric of the park and to prevent land managers from uniting and influencing the management strategies of the area. A mountain range that runs through the centre of the park, and divides the park into five different municipal authorities, was not mentioned by the land managers, even though it is a significant natural barrier that could limit contact between land managers.

The scientists identified research funding organizations as the predominant influence on their work and research direction. Organizational bodies that determine policy and regulation at the national and international levels were also mentioned as key actors. The Laird and agricultural organizations representing landowners, as well as park officials, were mentioned as actors that are easier to work with than many of the individual land managers.

The scientists reference to the local circle was typically in the context of bio-diversity conservation, but their view of local biodiversity issues was filtered through regional and global biodiversity requirements and at the long term.

3.2 Areas of concern

Using a combination of the findings from the interviews with the land managers and the scientists, together with the findings from the questionnaires, we organized the perceptions and attitudes expressed into five key themes (Fig. 2):

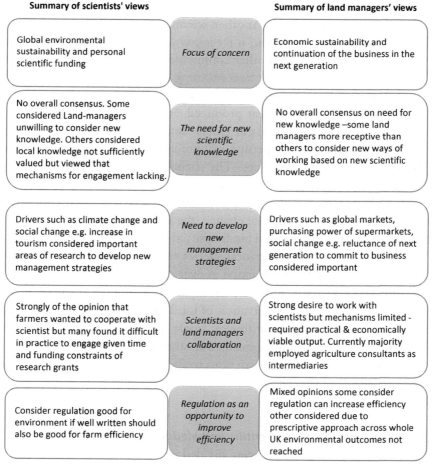

Summary of scientists' views		Summary of land managers' views
Global environmental sustainability and personal scientific funding	*Focus of concern*	Economic sustainability and continuation of the business in the next generation
No overall consensus. Some considered Land-managers unwilling to consider new knowledge. Others considered local knowledge not sufficiently valued but viewed that mechanisms for engagement lacking.	*The need for new scientific knowledge*	No overall consensus on need for new knowledge –some land managers more receptive than others to consider new ways of working based on new scientific knowledge
Drivers such as climate change and social change e.g. increase in tourism considered important areas of research to develop new management strategies	*Need to develop new management strategies*	Drivers such as global markets, purchasing power of supermarkets, social change e.g. reluctance of next generation to commit to business considered important
Strongly of the opinion that farmers wanted to cooperate with scientist but many found it difficult in practice to engage given time and funding constraints of research grants	*Scientists and land managers collaboration*	Strong desire to work with scientists but mechanisms limited - required practical & economically viable output. Currently majority employed agriculture consultants as intermediaries
Consider regulation good for environment if well written should also be good for farm efficiency	*Regulation as an opportunity to improve efficiency*	Mixed opinions some consider regulation can increase efficiency other considered due to prescriptive approach across whole UK environmental outcomes not reached

Fig. 2 Five areas of concern voiced by land managers and scientists related to sustainable knowledge-building.

3.2.1 Focus of concern

The concerns raised by all the land managers primarily dealt with the profits of their business and local issues, such as the social concerns about the Community network and the (lack of) continuity for future generations.

In contrast the predominant focus of scientists was on sustainability at the global rather than the local scale. Scientists rationalize these concerns around two factors:

(1) Funding—The source of funding for their research is typically from national and international funding bodies who are interested in global environmental sustainability. One of the scientists worded the difference in focus between scientists and land managers as follows: "*The government is interested in preserving the environment. Farmers are interested in conserving habitats*" (S4). Another scientist stated: "*The focus today [of EU and the UK government] is the impact [of land use] on the environment, not just GDP*" (S11). Another commented that "*the government concerns about the level of carbon emissions because of sheep diseases*" (S1), indicating that sheep diseases are responsible for increased emissions of greenhouse gas per kilo of sheep meat produced. S3 added that the problem of nitrogen emissions is a major global issue and it is caused, in part, by farming methods of dispersing manure and excessive use of fertilizers.

(2) Sustainability and resilience—The scientists focused on the sustainability of socio-ecological systems and recognized the need for data covering a large temporal and spatial scale. While this is obvious for questions of air pollution and climate change, the scientists considered it important when preserving local biodiversity and nature. One of the scientists explained that protecting local species required national management rather than isolated local action as everything is interconnected.

This difference in motivation and focus can sometimes lead to conflicted goals. For example, the conservation of local species may require the allocation of protected areas within a single land unit (farm or estate). This can be a sensitive point, explained another scientist, because land managers are afraid of potential restrictions to their business: "*land managers are sometimes afraid to cooperate with scientists for fear of finding a rare species in their land*" (S9).

3.2.2 The need for new scientific knowledge

Land managers and scientists were divided in their opinions regarding the need for new scientific knowledge. While most land managers did not raise the need for scientific advice, some of the scientists felt that it is necessary to raise awareness among land managers concerning problematic land management practices. The cohort of scientists questioned considered this necessary in order to convince land managers to work with scientists. At least three of the scientists emphasized the preference of farmers to continue using traditional methods instead of adopting what the scientists viewed as new and more appropriate methods, e.g., using precision-farming techniques or testing for parasite burden rather than prophylactic treatment of an entire flock. Two scientists noted that current agricultural methods in the park are considered "traditional," even though they only evolved into their present form after World War II. In addition, several scientists claimed that in many cases older land managers were in no hurry to transfer the management of the land to the younger generation but were also more reluctant to introduce management changes. Two of the scientists stressed the lack of availability of long-term data to the public, scientists and to decision makers. There was also a claim made by many of the scientists that since scientific publications are usually characterized by detailed methods and findings which require scientific background to understand them, scientific knowledge should be mediated to land managers through a summary of conclusions and recommendations, so that they could use it; a translation step was required.

Two survey questions focussed on the need for scientific advice: the first questioned whether the land managers obtained sufficient scientific advice from the government (Fig. 3A) and the second asked more directly if the land managers needed more scientific advice to improve their income from the land (Fig. 3B). Around a quarter of the land managers were uncertain and either did not answer or offered no opinion in response to both questions. While 44% of land managers felt that they received insufficient scientific advice from the government, only 29% of scientists reported that they considered land managers were lacking scientific advice from the government. While in response to the second question a similar proportion of both groups considered that land managers needed more scientific knowledge in order to improve the income of their farms (44% and 50% for land managers and scientists, respectively); the land managers more strongly agreed with this statement (Fig. 3B). Such a difference might indicate that land managers are more open to considering scientific knowledge than those scientists that were questioned had supposed.

The majority of land managers remarked that they wanted to cooperate with scientists to create new knowledge (Fig. 4), to improve soil fertility

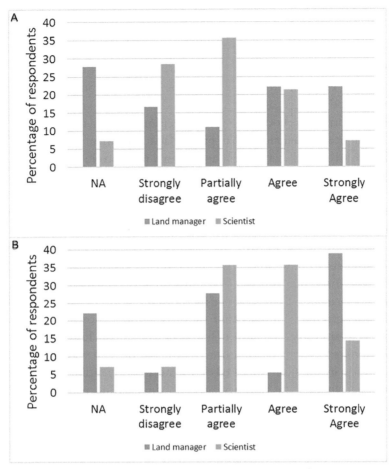

Fig. 3 Percentage of land managers and scientists reporting on a four-point Likert scale with additional "no opinion" or no answer option (NA). The graph shows their response to the statements: (A) Land managers do not get enough scientific advice from the government and (B) Land managers need more scientific knowledge in order to improve farm income ($n = 18$ and 15 for land managers and scientists, respectively).

(94%), reduce crop and livestock diseases (83%) and develop new crops/livestock breeds (67%). During the interviews, some land managers indicated a need to develop breeds of cattle and sheep that will withstand climate change, in particular cold weather. Scientists mirrored the opinion of land managers although fewer considered that farmers desired new knowledge on soil fertility (50% compared with 94%).

Three of the scientists we interviewed proposed "demonstration farms" as a solution to the lack of exposure of land managers to innovations and

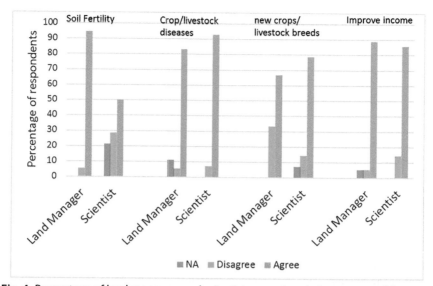

Fig. 4 Percentage of land managers and scientists reporting their agreement/disagreement or "no opinion" (NA) to statements related to their desire for new knowledge to improve soil fertility, combat crop/livestock diseases, develop new crops/livestock breed or improve income.

science-based solutions for land management. Yet, only a few scientists mentioned the economic barriers facing farmers who seek to adopt innovative technologies or management practices, as one of the scientists put it: "*in order for a farmer to adopt new technologies, there must be a convincing factual basis for their effectiveness. This is due to the high costs of change*" (S4).

The result of this perceived lack of translation to and understanding of scientific knowledge by land managers was reflected to some extent by the land managers. Several land managers stated a preference for receiving new scientific information from agricultural consultants. The advice of agricultural consultants was perceived by the land managers as more appropriate to their local problems and more sensitive economically. This perception was especially strong for land managers who had had prolonged relationships with an agricultural consultant, who knew the land manager's land well.

3.2.3 Collaboration between scientists and land managers
Among the land managers and scientists, there was general agreement on the value of cooperation, to formulate both suitable strategies and methods

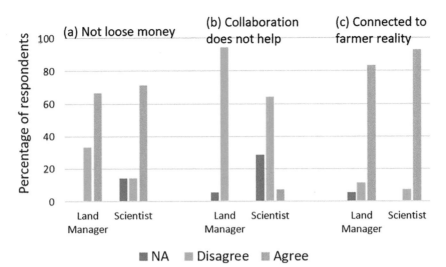

Fig. 5 Percentage of land managers and scientists reporting their agreement/disagreement or "no opinion" (NA) to statements related to their desire to cooperate with scientists; statement (A) willing to cooperate with scientists only if assured not to lose money, (B) cooperation with scientists does not help to improve the farm's income and (C) scientists must be more connected to the farmers reality.

for sustainable land management (Figs. 3 and 4). They were also in general agreement that collaboration should not result in a loss of income (Fig. 5A). Interestingly, a third of land managers were willing to take a calculated risk and participate in experiments that could result in a lower yield than they would normally achieve. They considered the risk worthwhile if they believed that the experiment would result in knowledge that would ultimately improve farm income. Land managers strongly believed (94%) that cooperation with scientists could help improve farm incomes (Fig. 5B), while scientists were less certain that farmers acknowledged their ability to help (64%). Both land managers (83%) and scientists (93%) considered the need for scientists to be more connected to the reality of farming in Scotland to be important.

Several considerations regarding the effectiveness of collaborating with scientists were raised by land managers. A common theme was related to the importance of local and practical knowledge. They noted that:

- Farmers have a long acquaintance with the land (LM3, LM12, LM17)
- It is important to integrate local and long-term practical knowledge into the management of the land (LM1, LM3, LM12)

- It is important to integrate local and long-term knowledge into land development programs (LM3)
- The government considers residents unable to manage the park and therefore, it is currently managed by urban people (LM1).

Scientists were asked with two specific questions about their belief in the role that land managers have to create knowledge on sustainable land management in Scotland (Table 3). All scientists interviewed reported that they considered farmer knowledge to be vital, with the majority (79%) reporting that they strongly agreed with the statement. However only two-thirds (64%) considered that it was necessary to co-produce knowledge with farmers, with around one-third indicating that they currently do not do this.

Demand for transdisciplinary studies, based on collaboration between researchers and stakeholders, is one of the prerequisite conditions of many of the research funding streams, e.g., European H2020 Responsible Research and Innovation (Holzer et al., 2018). However, scientists noted that sometimes it was difficult to engage land managers in funded research. It was noted that in many cases land managers do not attend collaborative meetings. According to the scientists, this is because they do not see any personal economic gain and perceive the meetings as a waste of time. However, the response of farmers to the questions related to economic gain when collaborating with scientists (Fig. 5) would suggest this is not the main reason. Some scientists admitted that they preferred to coordinate their studies

Table 3 Percentage of scientists reporting their agreement or disagreement to statements related to the role land managers should play in creating sustainable land management knowledge in Scotland and the role land managers currently have in their individual research portfolios.

Question	Strongly agree	Agree	Partially agree	Strongly disagree
I believe that farmer knowledge is vital to sustain farming in Scotland	79	21	0	0
Co-produced solution involving scientist and farmers is vital to sustain farming in Scotland	64	36	0	0
Most of my research is co-production of knowledge with farmers	0	21	50	29

with the Laird (the land owner) rather than many individual land managers, in order to save time. Similarly, most scientists testified that they usually work with farming organizations rather than the local land manager. Differences in language, sources of economic gain and of culture may therefore form a barrier to collaboration between local land managers and scientists.

3.2.4 Need to develop new management strategies

Land managers described social phenomena that could change the local SES, such as the sale of land to neighbours, due to the reluctance of the younger generation to become land managers, or new entrants being unable to raise sufficient capital to compete in the land market. This issue was raised in many of the land manager interviews (74% reported they agree or partially agree with the statement "The younger generation is not interested in agriculture"). Some land managers admitted that they encouraged their children to leave because of the low income in farming. However, one land manager told us that: "*today the situation is reversed because other jobs are not profitable*" (LM1). The reluctance of the younger generation to return to the land and manage the business was raised also in scientists' interviewees. However, 29% of scientists declined to answer the question ("The younger generation is not interested in agriculture"), indicating they were unaware of the situation (65% agreed and 7% disagreed). Another ongoing social phenomenon is the sale of land to urban people, who earn income outside of farming and therefore are not concerned about optimized land management (LM6, LM17). One of the major changes in land management noted in the interviews is the incorporation of more diverse land uses on a single unit, such as forestry, tourism, raising sheep for meat and hunting. Some of the scientists mentioned the changing trends in land management as requiring changes in management methods. For instance, "The creation of large farms with different land uses requires adjustment of management methods and strategies" (S10). This point was raised in the context of SES processes, such as climate change and social change. Scientist 1 emphasized the significant role of gamekeepers in managing the land; a role that does not receive enough attention by regulators, in her opinion. Here again, the particular focus of the scientist was on sustainability.

In comparison to the scientists, land managers emphasized market forces as the main reason to change their land management strategies. They stated that opening the local market to the global economy creates competition and lowers the price of meat, for example. They find it difficult to meet this competitive price because of land capability, climate and costs associated with livestock health and environmental regulations. Thirty six percent of the scientists preferred not to answer the question related to the impact of the wider economy on land managers (Fig. 6), while 50% of land managers agreed with the statement that the "Scottish economy in general is the most important obstacle to agriculture in Scotland" compared with 14% of scientists.

The economic uncertainty was also related to the withdrawal from the European Union (Brexit). Land managers who combine tourism on their land claimed that: *"the tourism industry"* ... *"fluctuates with the national and global economic situation. Therefore, it is necessary to combine additional types of income"* (LM4). The pronounced power of the supermarkets, as intermediaries in the sale of agricultural produce, was also mentioned as a major constraint for the land managers. One of the land managers claimed that *"the supermarkets pass the impact of the global economy on farmers and there is no strong association of farmers, which can regulate their power. In addition, there is not enough public support for local and organic economy. People stick to their values, as long as they do not cost them too much money"* (LM1).

In general, scientists and land managers agree that the future of the local economy is uncertain (Fig. 7), because of Brexit and climate and social changes. Similarly, land managers and scientists agreed (67% and 64%, respectively) that "Changing climate threatens agriculture" (Fig. 8).

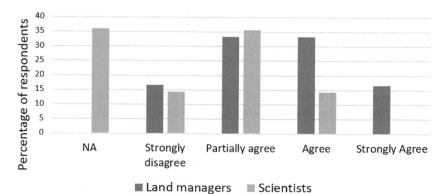

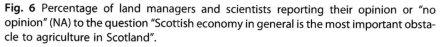

Fig. 6 Percentage of land managers and scientists reporting their opinion or "no opinion" (NA) to the question "Scottish economy in general is the most important obstacle to agriculture in Scotland".

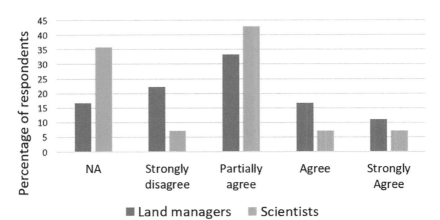

Fig. 7 Percentage of land managers and scientists reporting their opinion or "no opinion" (NA) to the statement: "The crisis in agriculture is temporary and the situation will improve in the coming years".

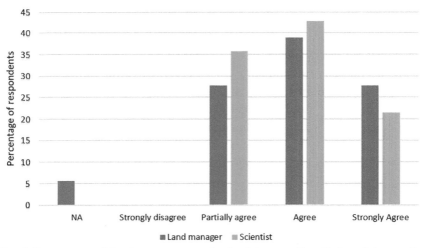

Fig. 8 Percentage of land managers and scientists reporting their opinion or "no opinion" (NA) to the statement: "Changing climate threaten agriculture".

3.2.5 Regulation as an opportunity to improve efficiency

The impact of environmental regulation in Scotland was a major issue in all interviews with both land managers and scientists. All land managers at least partially agreed with the statement that "Regulations are major problems for farming" (compared with 71% of scientists). The interviews revealed that

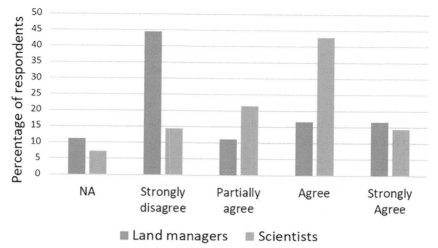

Fig. 9 Percentage of land managers and scientists reporting their opinion or "no opinion" (NA) to the statement: "Environmental regulation can improve my business".

land managers and scientists agree that regulations are designed to protect the environment and not the economic sustainability of the land managers. Some 44% of land managers considered environmental regulations to be detrimental to their economic sustainability as indicated by their response to the statement "Environmental regulation can improve my businesses" compared with 79% of scientists who at least partially agreed with the statement (Fig. 9).

Although there was no consensus on how regulation affects land management, the interviews did provide some interesting insights. Many of the scientists regarded regulation as an opportunity for economic improvement, which calls for cooperation between land managers and scientists. However, they consider that the tone of the regulatory message should be changed. One of the scientists remarked: "*The message today to the farmers: change the agricultural methods to comply with regulation. Instead, the message should be: change the methods because it is more economical to recycle cow manure and produce energy from it.*" (S3). More than half of the land managers (51%) also did not reject the possibility that environmental regulations could improve their businesses. However, some land managers complained that the regulations are confusing, and they find it difficult to understand how to comply with them. In response, scientists told us that there is an interest in researching the barriers that land managers face, for example, adapting agriculture methods to the regulations. Another issue raised by both land managers and scientists was that the regulations should be modified to the local environmental conditions and the seasons within the United Kingdom. The lack of compatibility to the unique conditions of the north and south of Scotland causes problems and losses for land managers, and in some cases harms the environment. These views would emphasize the need for local and collaborative research. One of scientists claimed that: *"Farmers are not trying to increase efficiency but to reduce losses"* (S4).

4. Discussion

Transdisciplinary research, although widely accepted as an appropriate way to build the knowledge required for sustainable land management (Glass et al., 2013; Holzer et al., 2018; Jax et al., 2018; Moore, 2013; van der Hel, 2016), is fraught with difficulties (Angelstam et al., 2018; Teschner et al., 2017; Thompson et al., 2017). The challenge of collaboration and co-development by land managers, scientists and agricultural consultants for building resilient SESs is very relevant in Scotland, in view of the recent report of the Scottish Parliament that shows agriculture is actually a risk-management business (Thomas, 2018). The findings from our interviews with the CNP's land managers and with scientists who are conducting research in the park, revealed differences in the perception of both groups regarding the local social–ecological system (SES). These differences can explain some of the difficulties they have had in cooperating, but they also suggest new areas and opportunities for collaboration. In the following sections we summarize the differences in SES perception. We will then suggest a possible way for fruitful cooperation and transdisciplinary research by land managers and scientists.

4.1 Perceptions of the SES

The socio-ecological system of the Cairngorms National Park (CNP), as portrayed by the land manager and scientist groups, was complex and open. However, each group described a system that included different sets of critical actors and unique networks of socio-ecological interaction. The analysis of the actor-maps indicated that the land managers generally referred to the land under their management, their families and the farms nearby as the SES. Interestingly, agricultural consultants were mentioned in most cases as part of this internal system and not as an external factor. Scientists, in contrast, referred to the entire CNP as the SES and sometimes placed the park as part of the rural SES of Scotland or United Kingdom. The social-system's structure and function described was also different in both groups. While the land managers emphasized economic factors and regulators, the scientists typically mentioned research organizations, foundations and government bodies as critical actors. The way in which the two groups referred to the boundaries of the system dictated their attitude to the various actors, as internal or external to the SES. This point is important because people generally feel more able to influence actors who are part of their social system (Burns, 2007). The reference to the ecological part of the SES was also different in both groups. While the land managers hardly mentioned it, or they referred to biodiversity in the context of oppressive environmental regulation, for scientists the ecology was a major issue, both in the local and global context.

4.2 Barriers for sustainable land management

It was apparent from the high percentage of the choice, "no opinion" or "partially agree" in Figs 6–8, and more widely across the interviews, that there is a lack of confidence in the resilience of the current social-ecological system (SES) of Cairngorms National Park. Land managers and scientists were unsure how the CNP's SES will be affected by local and global processes. The complexity of the system and the many forces influencing it that are undergoing change—external and global (e.g., global economy and climate change); national and international (e.g., government policy and environmental regulations of the European Union); and regional and local (e.g., changes in consumers' taste, sale and consolidation of farms, the uncertain continuity by the young generation)—make it very difficult to predict the expected directions of change and the long-term influences of land management decisions, and therefore to plan future land management within the park. Indeed, local SESs are subject to the effects of environmental, economic, social and political processes across large scales of time and space (Costanza et al., 2014; Dearing et al., 2010; Ericksen, 2008; Folke et al., 2010; Lambin et al., 2001; Plummer and Armitage, 2007). In Scotland, in the autumn of 2017, the approaching Brexit added to this sense of uncertainty.

Despite this apparent consensus, our findings indicate differences between the two groups in relation to barriers for sustainability and the desired future. Land managers emphasized short-term economic and social sustainability considerations. The desired enterprises of the farm were considered (e.g., sheep for meat, afforestation, games and tourism), as well as efficient agricultural methods and the need to cope with intermediaries and compete with world markets. In contrast, the scientists stressed global and long-term environmental sustainability. Although none of the scientists interviewed suggested conversion of the agro-ecological system within the national park into a natural-ecological system as a way to increase carbon sequestration at the national level, for example, the scientists did note the harmful effect of some conventional agricultural practices on the local and global environment, and emphasized the importance of broad-scale environmental regulation. Similar broad-scale environmental regulation approaches can be seen in Heck et al. (2018), who advocated the development of a global model of land management, to optimize carbon fixation, food production and ecosystem preservation. Likewise, Hardaker (2018) points to low afforestation rates in England, compared to other parts of Europe, as a motivation to convert grazing lands into forests, rather than considering what might be the specific local need. Similar global, environmental perspectives have also been expressed by other researchers (e.g., Costanza et al., 2014; Jax et al., 2018; Magliocca et al., 2018; Mirtl et al., 2018; Ostrom, 2010; van der Hel, 2016). Land managers saw environmental regulation as a confusing and burdensome factor, however, like the scientists there was some agreement among land managers that environmental regulation might serve as a mechanism, an opportunity, for boosting efficiency (Fig. 9). Variability in funding, whether via agricultural subsidies to ensure land manager economic income (Bateman and Balmford, 2018; Grant, 2016) or the funding of scientific research (Cressey, 2016), is a aspect of this uncertainty shared by both groupings. That the scientists and the land managers shared a sense of uncertainty regarding the future did appear to foster a desire for cooperative working between these two groups on the resilience of the Cairngorms National Park SES.

4.3 What is the required knowledge?

Our interviews revealed that the level and type of knowledge required to build a resilient socio-ecological system varied, and this depended on the different perspectives regarding the entire system held by land managers and scientists. The land managers argued that proven answers to practical agricultural and economic questions are needed; the type of knowledge required is essentially practical in nature. The scientists, on the other hand, stated that they typically evaluate the long-term effects of agricultural practices and policy on the environment (e.g., air and water quality) and on biodiversity. They develop generic knowledge, often in collaboration with or in comparison to other global SES systems (e.g., IPBES papers). To obtain reliable conclusions, relatively long-term research is required and the answers obtained from any study may not be practically applicable. Such delays are frustrating for those who seek an immediate but reliable answer. Land manager preferences for being advised by agricultural consultants rather than scientists (Fig. 4 and statements from the interviews) are therefore understandable (see also Teschner et al., 2017). Another constraint revealed by the interviews was the detailed and formal way that findings are reported in the scientific literature. Papers are often overloaded with information and do not communicate specific, practical answers. Thus, many scientists who participated in the study voiced the need for simplification and adaptation of scientific knowledge to the needs of land managers—a task of technology transfer that has challenged many researchers (e.g., Jax et al., 2018; Mielke et al., 2016; Wise et al., 2014). One conclusion from this might be that scientists should consider working with agricultural consultants rather than directly with land managers. Professional consultants may more reliably express the problem and formulate the required, practical solution in a way that is transferable to both scientists and local land managers. However, some of the scientists also noted that the younger generation of land managers typically have academic degrees in agriculture and are capable of understanding and applying scientific knowledge directly, which might indicate that mixed approaches, both directly with land managers and via consultants, may be most suitable in the future.

4.4 Readiness to cooperate

When we embarked on our research, we anticipated that land managers would be reluctant to meet and talk to us, as scientists. To improve our chances of conducting interviews, we timed our field work for the beginning of autumn when farmers in Scotland are typically less busy on their land and planned to meet them at places of their choosing. To our surprise, most of the land managers we approached agreed to be interviewed and invited us into their homes. The interviews were prolonged, and the general impression was that the interviewees were happy to express their opinion on the situation—some remarking that they are seldom asked their opinion. The readiness of land managers and scientists to cooperate was also expressed in the response to the survey questions on this subject (Fig. 5).

The contrasting perceptions of the socio-ecological system of the CNP may complicate and limit cooperation between scientists and land managers, but, as was noted by some of the interviewees, these differences may also be an opportunity to expand the influence of the stakeholders within the system. For example, inter-relations between land managers exchanging information may increase the impact of scientific discoveries throughout the park area. Likewise, these different views may be an opportunity to promote transdisciplinary exchange between these two diverse worlds of knowledge (land management and scientific), as demonstrated in Section 4.4. There are also controversial issues that can serve as a basis for collaborative work. For example, EU environmental regulation, which was mentioned earlier as a focal issue for both land managers and scientists, may create an opportunity for cooperation (see Section 3.2.5).

4.5 Formulating a future-vision as a long-term and transdisciplinary process

One common way of dealing with the need to make decisions in conditions of uncertainty is by building scenarios (Priess et al., 2018). Scenario building, based on relevant factors and theory, examines how choosing certain practices will affect the complex system. The advantage of this approach is that it holds the possibility of identifying those parameters that are key both for long-term resilience and for monitoring changes in the socio-ecological system. However, scenario building does have a weakness in that it necessarily simplifies the system and may therefore exclude some vital parameters. Hauck et al. (2019) demonstrated how including policy and regulatory factors in commonly used, scenario-based models can significantly change land use outcomes.

The importance of including the perspective of stakeholders in the future-vision has been demonstrated by Brown et al. (2018), who also emphasized the possibility that stakeholders might modify their behaviour in response to changing circumstances in SES. Indeed, the complexity of the social system can make it difficult to identify social functions that may impede adoption of desired actions, especially where those stakeholders are not partners in the shaping of the future-vision (Brown et al., 2018; Bruns and Worsley, 2015; Canova et al., 2019; Shaw et al., 2009). Fostering the resilience of social-ecological systems and binding involvement of stakeholders in planning the future of the region is central to this approach. This goal can be reached by conducting transdisciplinary research

(Bai et al., 2016; Fazey et al., 2014; Holzer et al., 2018; Mirtl et al., 2018; Plummer et al., 2017; Popa et al., 2015; van der Hel, 2018).

Building a future-vision is a process of synthesis, amalgamating practical knowledge, motivations and aspirations of stakeholders when picturing a desirable future. Future-visions have been found to have a positive impact on the motivation of organizations (Levin, 2000) as well as for private stakeholders to adopt changes (Brown et al., 2018). In order to establish fruitful transdisciplinary research within the Cairngorms National Park, we therefore propose a participatory action-research approach, which is based on cycles of research that include: (a) identifying a problem; (b) analysing the field and formulating a research question; (c) raising a solution and applying it while meticulously collecting data to; and, (d) examining the effect of the action on the research field. Analysis and reflection on the results of the research cycle may then lead to reformulation of the research question and another cycle of action-research (Fig. 10). The action research circles in effect form an ongoing spiral of research moving toward a potentially appropriate solution for the problem under study. Action researchers around the world have been using this approach for about 70 years to generate social-environmental change (see, for example, Burns, 2007; Zuber-Skerritt, 2012). Here we propose to conduct participatory action-research as a coupled spiral process. Land managers, scientist, as well as professional advisors in the relevant fields collaborate and express their multiple points of view on formulating ongoing challenges, applying and adapting the future-vision to the changing conditions potentially improve the resilience of the social-ecological system (SES) of the Cairngorms National Park (Fig. 10).

Participatory action-research is transdisciplinary (Avriel-Avni et al., 2017). It starts with real-world, local problems, but is also a problem-based research approach that suggests generic solutions and not "one solution for each situation" (Bai et al., 2016; Jax et al., 2018; Scott, 2011). It can also produce more acceptable local change because land managers bring into this collaboration their deep and long-term acquaintance with the (bio-physical and social) locality and its real-world problems.

4.6 Conclusions and recommendations

The results of this study indicate significant differences between land managers and scientists in the perception of the social-ecological system. Differences were found in defining the boundaries of the coupled system, including what is considered part of it and what is considered an external

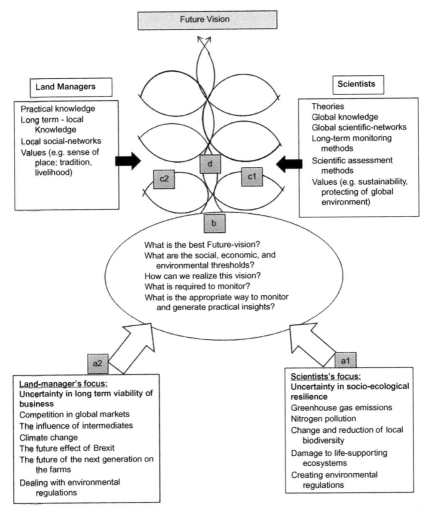

Fig. 10 Formulating future-vision through participatory action-research. The research is conducted through coupled spirals. The starting points (a1, a2) are dissatisfaction with the current situation. Formulation of a common research question and a possible solution (b) lead to implementation of the solution, while collecting data on its effect (c1, c2). The next step (d) is analysis of the findings, reflection, refinement or change of the research question and improvement or modification of the solution. The spiral continues to exist until both sides are satisfied with the new situation.

driver. While this paper focusses on the relationship between land managers and scientists the role of other relevant factors such as policy makers, regulatory agency staff and consumers will also influence outcomes. Although both groups agree that the system involves multi-actors, their definitions of the relevant social network and the main problems to be solved are different. Land managers naturally focus on challenges on their land and are typically less concerned about regional processes, while scientists focus on land management challenges in the context of local and global sustainability.

Despite the complexity of the SES, land managers are required to make decisions even though they are uncertain about the future trends of various processes in the local, national and global systems. The decisions that land managers make, while focussed on local economic efficiency, can have a major impact on the resilience of the entire SES.

Conducting long-term collaborations with researchers from different disciplines, i.e., transdisciplinary research, may be a suitable way to foster resilience by providing targeted knowledge suitable for local situations but with a wider perspective. Indeed, our findings indicate that land managers and scientists are willing to collaborate to formulate solutions. However, because land managers focus on managing their lands and solving local problems, it is necessary to consider a suitable framework for such transdisciplinary research.

One consistent aspect of transdisciplinarity research is the theme of collaboration. In the case of CNP, as in many agricultural SES worldwide, the challenge of weaving together a future-vision would benefit from a regional perspective instead of focussing on individual land managers and farms. Indeed, the multiplicity of actors in the SES has a significant impact on the structure and function of the system and this must be taken into account in any future-vision for an SES system. We suggest that building a joint future-vision would be the basis for fruitful collaboration.

The second aspect of the transdisciplinary research is the collaborative framework. The practical need for regional and long-term thinking could be operationalized through the Monitor Farm Programme's model in Scotland (ADAS, 2008) at the landscape scale. Monitor Farms are based on working farms, and thus represent the complexity of the system and the involvement of the various actors. Presently, single monitor farms in Scotland focus on the whole farm approach and while still maintaining sector specialisms, help Scottish farmers to develop new farm business management practices, resulting in improved agricultural efficiency and environmental management, and mitigating climate change on their farm. Currently, one monitor farm functions for a period of 3–5 years (typically 10 across Scotland at any one time; https://www.monitorfarms.co.uk/). Establishing several neighbouring monitor farms in a region such as the Cairngorm National Park, would create a demonstration network and would encourage a regional perspective. Land managers and scientists would be able to continually monitor and demonstrate the adoption of actions from the future-vision, along with their efficiency and impact on the resilience of the whole social-ecological system.

Acknowledgements

The authors are very grateful to all interviewees who took part in this study, who provided the original information contained in this paper. The goals of the study could not have been achieved without their active, willing participation. The authors are indebted to them for sharing their knowledge, time and wisdom with us. This study was part funded by the eLTER program European Union Horizon 2020 Grant Number 654359 "European Long-Term Ecosystem and Socio-Ecological Research Infrastructure—eLTER" and by Scottish National Heritage via the UK Environmental Change Network, although the views expressed are those of the author alone. The authors are also indebted to the two anonymous reviewers whose insightful comments improved the quality of this paper.

References

ADAS, 2008. An Investigation Into the Role and Effectiveness of Scottish Monitor Farms in Improving the Sustainability and Profitability of Participating Farm Business and Disseminating the Results to Influence the Wider Farming Community. Report produced for the Scottish Government Rural Directorate.

Angelstam, P., Manton, M., Elbakidze, M., Sijtsma, F., Adamescu, M.C., Avni, N., Beja, P., Bezak, P., Zyablikova, I., Cruz, F., Bretagnolle, V., 2018. LTSER platforms as a place-based transdisciplinary research infrastructure: learning landscape approach through evaluation. Landsc. Ecol. 1–24. https://doi.org/10.1007/s10980-018-0737-6.

Argyris, C., Schön, D.A., 1974. Theory in Practice: Increasing Professional Effectiveness. Jossey-Bass, San Francisco.

Avriel-Avni, N., Holzer, J.M., Shachak, M., Orenstein, D.E., Groner, E.E., 2017. Using transdisciplinary action research toward sustainable management of vineyard management and tourism in the Negev highlands. In: Mapotse, T. (Ed.), Cross-Disciplinary Approaches to Action Research and Action Learning. IGI Global, Hershey, PA, pp. 215–226.

Bai, X., Van Der Leeuw, S., O'Brien, K., Berkhout, F., Biermann, F., Brondizio, E.S., Cudennec, C., Dearing, J., Duraiappah, A., Glaser, M., Revkin, A., 2016. Plausible and desirable futures in the Anthropocene: a new research agenda. Glob. Environ. Chang. 39, 351–362.

Barlow, J., Ewers, R.M., Anderson, L., Aragao, L.E., Baker, T.R., Boyd, E., Feldpausch, T.R., Gloor, E., Hall, A., Malhi, Y., Milliken, W., 2011. Using learning networks to understand complex systems: a case study of biological, geophysical and social research in the amazon. Biol. Rev. 86 (2), 457–474. https://doi: 10.1111/j.1469-185X.2010.00155.x.

Barnes, M.L., Bodin, Ö., Guerrero, A.M., McAllister, R.J., Alexander, S.M., Robins, G., 2017. The social structural foundations of adaptation and transformation in social-ecological systems. Ecol. Soc. 22 (4), 16. https://doi.org/10.5751/ES-09769-220416.

Bateman, I.J., Balmford, B., 2018. Public funding for public goods: a post-Brexit perspective on principles for agricultural policy. Land Use Policy 79, 293–300.

Belmont Forum, 2011. The Belmont Challenge: A Global Environmental Research Mission for Sustainability. https://www.igfagcr.org/sites/default/files/documents/belmont-challenge-white-paper.pd.

Bohnet, I., Smith, D.M., 2007. Planning future landscapes in the wet tropics of Australia: a social–ecological framework. Landsc. Urban Plan. 80 (1–2), 137–152.

Brown, C., Holzhauer, S., Metzger, M.J., Paterson, J.S., Rounsevell, M., 2018. Land managers' behaviours modulate pathways to visions of future land systems. Reg. Environ. Chang. 18 (3), 831–845.

Bruns, D., Worsley, S., 2015. Navigating Complexity in International Development. Facilitating Sustainable Change at Scale. Practical Action Publishing, Rugby, UK.

Burns, D., 2007. Systemic Action Research: A Strategy for Whole System Change. Policy Press.

Cairngorms National Park Authority, 2015. Cairngorms National Park Research Strategy 2014–2017. https://cairngorms.co.uk/authority/publication/352/ (accessed 2nd July, 2018).

Canova, M.A., Lapola, D.M., Pinho, P., Dick, J., Patricio, G.B., Priess, J.A., 2019. Different ecosystem services, same (dis)satisfaction with compensation: a critical comparison between farmers' perception in Scotland and Brazil. Ecosyst. Serv. 35, 164–172. https://doi.org/10.1016/J.ECOSER.2018.10.005.

Carmen, E., Watt, A., Young, J., 2018. Arguing for biodiversity in practice: a case study from the UK. Biodivers. Conserv. 27 (7), 1599–1617.

Costanza, R., Wainger, L., Folke, C., Mäler, K.G., 1993. Modeling complex ecological economic systems: toward an evolutionary, dynamic understanding of people and nature. Bioscience 43 (8), 545–555.

Costanza, R., de Groot, R., Sutton, P., van der Ploeg, S., Anderson, S.J., Kubiszewski, I., Farber, S., Turner, R.K., 2014. Changes in the global value of ecosystem services. Glob. Environ. Chang. 26 (1), 152–158.

Cressey, D., 2016. Scientists say 'no' to UK exit from Europe in Nature poll. Nature 531 (7596), 559.

Dearing, J.A., Braimoh, A.K., Reenberg, A., Turner, B.L., Van der Leeuw, S., 2010. Complex land systems: the need for long time perspectives to assess their future. Ecol. Soc. 15 (4), 21.

Deary, H., Warren, C.R., 2017. Divergent visions of wildness and naturalness in a storied landscape: practices and discourses of rewilding in Scotland's wild places. J. Rural. Stud. 54, 211–222.

Díaz, S., Demissew, S., Carabias, J., Joly, C., Lonsdale, M., Ash, N., Larigauderie, A., Adhikari, J.R., Arico, S., Báldi, A., Bartuska, A., 2015. The IPBES Conceptual Framework—connecting nature and people. Curr. Opin. Environ. Sustain. 14, 1–16.

Dick, J., Verweij, P., Carmen, E., Rodela, R., Andrews, C., 2017. Testing the ecosystem service cascade framework and QUICKScan software tool in the context of land use planning in Glenlivet estate Scotland. Int. J. Biodivers. Sci. Ecosyst. Serv. Manag. 13 (2), 12–25.

Dick, J., Turkelboom, F., Woods, H., Iniesta-Arandia, I., Primmer, E., Saarela, S.R., Bezák, P., Mederly, P., Leone, M., Verheyden, W., Kelemen, E., 2018a. Stakeholders' perspectives on the operationalisation of the ecosystem service concept: results from 27 case studies. Ecosyst. Serv. 29, 552–565.

Dick, J., Orenstein, D.E., Holzer, J.M., Wohner, C., Achard, A.L., Andrews, C., Avriel-Avni, N., Beja, P., Blond, N., Cabello, J., Chen, C., 2018b. What is socio-ecological research delivering? A literature survey across 25 international LTSER platforms. Sci. Total Environ. 622, 1225–1240.

Ericksen, P.J., 2008. Conceptualizing food systems for global environmental change research. Glob. Environ. Chang. 18 (1), 234–245.

Evely, A.C., Fazey, I., Pinard, M., Lambin, X., 2008. The influence of philosophical perspectives in integrative research: a conservation case study in the Cairngorms National Park. Ecol. Soc. 13 (2), 52. [online] URL. http://www.ecologyandsociety.org/vol13/iss2/art52/.

Fazey, I., Bunse, L., Msika, J., Pinke, M., Preedy, K., Evely, A.C., Lambert, E., Hastings, E., Morris, S., Reed, M.S., 2014. Evaluating knowledge exchange in interdisciplinary and multi-stakeholder research. Glob. Environ. Chang. 25, 204–220.

Folke, C., 2006. Resilience: the emergence of a perspective for social–ecological systems analyses. Glob. Environ. Chang. 16, 253–267.

Folke, C., Carpenter, S., Walker, B., Scheffer, M., Chapin, T., Rockström, J., 2010. Resilience thinking: integrating resilience, adaptability and transformability. Ecol. Soc. 15 (4), 20.

Fowler, D., Coyle, M., Skiba, U., Sutton, M.A., Cape, J.N., Reis, S., Sheppard, L.J., Jenkins, A., Grizzetti, B., Galloway, J.N., Vitousek, P., 2013. The global nitrogen cycle in the twenty-first century. Philos. Trans. R. Soc. B 368 (1621), 20130164.

Glass, J., Price, M., Warren, C., Scott, A. (Eds.), 2013. Lairds, Land and Sustainability: Scottish Perspectives on Upland Management. Edinburgh University Press.

Grant, W., 2016. The challenges facing UK farmers from Brexit. EuroChoices 15 (2), 11–16.

Haasnoot, M., Kwakkel, J.H., Walker, W.E., ter Maat, J., 2013. Dynamic adaptive policy pathways: a method for crafting robust decisions for a deeply uncertain world. Glob. Environ. Chang. 23 (2), 485–498.

Haberl, H., Winiwarter, V., Andersson, K., Ayres, R.U., Boone, C., Castillo, A., Cunfer, G., Fischer-Kowalski, M., Freudenburg, W.R., Furman, E., Kaufmann, R., 2006. From LTER to LTSER: conceptualizing the socioeconomic dimension of long-term socio-ecological research. Ecol. Soc. 11 (2), 13. https://www.ecologyandsociety.org/vol11/iss2/art13/main.html.

Hardaker, A., 2018. Is forestry really more profitable than upland farming? A historic and present-day farm level economic comparison of upland sheep farming and forestry in the UK. Land Use Policy 71, 98–120.

Hauck, J., Schleyer, C., Priess, J.A., Veerkamp, C.J., Dunford, R., Alkemade, R., Berry, P., Primmer, E., Kok, M., Young, J., Haines-Young, R., 2019. Combining policy analyses, exploratory scenarios, and integrated modelling to assess land use policy options. Environ. Sci. Pol. 94, 202–210.

Haughton, A.J., Bond, A.J., Lovett, A.A., Dockerty, T., Sünnenberg, G., Clark, S.J., Bohan, D.A., et al., 2009. A novel, integrated approach to assessing social, economic and environmental implications of changing rural land-use: a case study of perennial biomass crops. J. Appl. Ecol. 46 (2), 315–322. https://doi.org/10.1111/j.1365-2664.2009.01623.x.

Heck, V., Hoff, H., Wirsenius, S., Meyer, C., Kreft, H., 2018. Land use options for staying within the planetary boundaries–synergies and trade-offs between global and local sustainability goals. Glob. Environ. Chang. 49, 73–84.

Hermans, C., Erickson, J., Noordewier, T., Sheldon, A., Kline, M., 2007. Collaborative environmental planning in river management: an application of multicriteria decision analysis in the white river watershed in Vermont. J. Environ. Manag. 84 (4), 534–546.

Holden, E., Lingered, K., Banister, D., 2014. Sustainable development: our common future revisited. Glob. Environ. Chang. 26, 130–139.

Holzer, J.M., Carmon, N., Orenstein, D.E., 2018. A methodology for evaluating transdisciplinary research on coupled socio-ecological systems. Ecol. Indic. 85, 808–819.

Janssen, M.A., Schoon, M.L., Ke, W., Börner, K., 2006. Scholarly networks on resilience, vulnerability and adaptation within the human dimensions of global environmental change. Glob. Environ. Chang. 16 (3), 240–252.

Jax, K., Furman, E., Saarikoski, H., Barton, D.N., Delbaere, B., Dick, J., Duke, G., Görg, C., Gómez-Baggethun, E., Harrison, P.A., Maes, J., Pérez-Soba, M., Saarela, S.-R., Turkelboom, J., van Dijk, F., Watt, A.D., 2018. Handling a messy world: lessons learned when trying to make the ecosystem services concept operational. Ecosyst. Serv. 29, 415–427.

Krasny, M.E., Roth, W.M., 2010. Environmental education for social–ecological system resilience: a perspective from activity theory. Environ. Educ. Res. 16 (5–6), 545–558.

Lambin, E.F., Turner, B.L., Geist, H.J., Agbola, S.B., Angelsen, A., Bruce, J.W., Coomes, O.T., Dirzo, R., Fischer, G., Folke, C., George, P., 2001. The causes of land-use and land-cover change: moving beyond the myths. Glob. Environ. Chang. 11 (4), 261–269.

Levin, I.M., 2000. Vision revisited: telling the story of the future. J. Appl. Behav. Sci. 36 (1), 91–107.

Löf, A., 2010. Exploring adaptability through learning layers and learning loops. Environ. Educ. Res. 16 (5–6), 529–543.

Magliocca, N.R., Ellis, E.C., Allington, G.R., De Bremond, A., Dell'Angelo, J., Mertz, O., Messerli, P., Meyfroidt, P., Seppelt, R., Verburg, P.H., 2018. Closing global knowledge gaps: producing generalized knowledge from case studies of social-ecological systems. Glob. Environ. Chang. 50, 1–14.

Mauser, W., Klepper, G., Rice, M., Schmalzbauer, B.S., Hackmann, H., Leemans, R., Moore, H., 2013. Transdisciplinary global change research: the co-creation of knowledge for sustainability. Curr. Opin. Environ. Sustain. 5 (3–4), 420–431.

MEA (Millennium Ecosystem Assessment), 2005. Ecosystems and Human Well-Being: A Framework for Assessment. The Millennium Ecosystem Assessment Series, Island Press, Washington, DC.

Mielke, J., Vermassen, H., Ellenbeck, S., Milan, B.F., Jaeger, C., 2016. Stakeholder involvement in sustainability science—a critical view. Energy Res. Soc. Sci. 17, 71–81.

Mirtl, M., Borer, E.T., Djukic, I., Forsius, M., Haubold, H., Hugo, W., Jourdan, J., Lindenmayer, D., McDowell, W.H., Muraoka, H., Orenstein, D.E., 2018. Genesis, goals and achievements of long-term ecological research at the global scale: a critical review of ILTER and future directions. Sci. Total Environ. 626, 1439–1462.

Moore, H., 2013. Transdisciplinary global change research: the co-creation of knowledge for sustainability. Curr. Opin. Environ. Sustain. 5 (3–4), 420–431.

Noy, C., 2008. Sampling knowledge: the hermeneutics of snowball sampling in qualitative research. Int. J. Soc. Res. Methodol. 11 (4), 327–344.

Ostrom, E., 2010. Polycentric systems for coping with collective action and global environmental change. Glob. Environ. Chang. 20 (4), 550–557. https://www.sciencedirect.com/science/article/pii/S0959378010000634.

Plummer, R., Armitage, D., 2007. A resilience-based framework for evaluating adaptive co-management: linking ecology, economics and society in a complex world. Ecol. Econ. 61 (1), 62–74. https://doi.org/10.1016/j.ecolecon.2006.09.025.

Plummer, R., Baird, J., Dzyundzyak, A., Armitage, D., Bodin, Ö., Schultz, L., 2017. Is adaptive co-management delivering? Examining relationships between collaboration, learning and outcomes in UNESCO biosphere reserves. Ecol. Econ. 140, 79–88.

Popa, F., Guillermin, M., Dedeurwaerdere, T., 2015. A pragmatist approach to transdisciplinarity in sustainability research: from complex systems theory to reflexive science. Futures 65, 45–56.

Priess, J.A., Hauck, J., Haines-Young, R., Alkemade, R., Mandryk, M., Veerkamp, C., Gyorgyi, B., Dunford, R., Berry, P., Harrison, P., Dick, J., 2018. New EU-scale environmental scenarios until 2050–scenario process and initial scenario applications. Ecosyst. Serv. 29, 542–551.

Ruhl, J.B., Kraft, S.E., Lant, C.L., 2007. The Law and Policy of Ecosystem Services. Island Press, Washington. http://books.google.co.il/books?id=VEF48vqc0zcC.

Scott, A., 2011. Beyond the conventional: meeting the challenges of landscape governance within the European landscape convention? J. Environ. Manag. 92 (10), 2754–2762.

Shaw, A., Sheppard, S., Burch, S., Flanders, D., Wiek, A., Carmichael, J., Robinson, J., Cohen, S., 2009. Making local futures tangible—synthesizing, downscaling, and visualizing climate change scenarios for participatory capacity building. Glob. Environ. Chang. 19 (4), 447–463.

Sim, J., Saunders, B., Waterfield, J., Kingstone, T., 2018. Can sample size in qualitative research be determined a priori? Int. J. Soc. Res. Methodol. 21 (5), 619–634.

Sutherland, L.A., 2010. Environmental grants and regulations in strategic farm business decision-making: a case study of attitudinal behaviour in Scotland. Land Use Policy 27, 415–423. https://doi.org/10.1016/j.landusepol.2009.06.003.

TEEB, 2010. In: Kumar, P. (Ed.), The Economics of Ecosystems and Biodiversity Ecological and Economic Foundations. Earthscan, London and Washington.

Teschner, N.A., Orenstein, D.E., Shapira, I., Keasar, T., 2017. Socio-ecological research and the transition toward sustainable agriculture. Int. J. Agric. Sustain. 15 (2), 99–101.

Thomas, G., 2018. Risk Management in Agriculture. The Scottish Parliament. Available at https://sp-bpr-en-prod-cdnep.azureedge.net/published/2018/7/4/Risk-management-in-agriculture/SB%2018-46.pdf [6 January 2019].

Thompson, M.A., Owen, S., Lindsay, J.M., Leonard, G.S., Cronin, S.J., 2017. Scientist and stakeholder perspectives of transdisciplinary research: early attitudes, expectations, and tensions. Environ. Sci. Policy 74, 30–39.

Turner, R.K., Georgiou, S., Fisher, B., 2008. Valuing Ecosystem Services: The Case of Multi-Functional Wetlands. Earthscan, London. http://books.google.co.il/books?id=ZacCAWBKHW8C.

Vaismoradi, M., Jones, J., Turunen, H., Snelgrove, S., 2016. Theme development in qualitative content analysis and thematic analysis. J. Nurs. Educ. Pract. 6 (5), 100–110.

van der Hel, S., 2016. New science for global sustainability? The institutionalisation of knowledge co-production in future earth. Environ. Sci. Policy 61, 165–175.

van der Hel, S., 2018. Science for change: a survey on the normative and political dimensions of global sustainability research. Glob. Environ. Chang. 52, 248–258.

Wamsler, C., 2017. Stakeholder involvement in strategic adaptation planning: transdisciplinarity and co-production at stake? Environ. Sci. Policy 75, 148–157.

Wise, R.M., Fazey, I., Smith, M.S., Park, S.E., Eakin, H.C., Van Garderen, E.A., Campbell, B., 2014. Reconceptualising adaptation to climate change as part of pathways of change and response. Glob. Environ. Chang. 28, 325–336.

Zuber-Skerritt, O. (Ed.), 2012. Action Research for Sustainable Development in a Turbulent World. Emerald Group Publishing, Bingley, UK.

Further reading

Allen, C., Birge, H., Angeler, D., Arnold, C., Chaffin, B., DeCaro, D., Garmestani, A., Gunderson, L., 2018. Quantifying uncertainty and trade-offs in resilience assessments. Ecol. Soc. 23 (1), 3.

Carlile, P.R., 2002. A pragmatic view of knowledge and boundaries: boundary objects in new product development. Organ. Sci. 13, 442–455.

Toomey, A.H., Markusson, N., Adams, E., Brockett, B., 2015. Inter- and Trans-Disciplinary Research: A Critical Perspective. Policy Brief. [Online]. Available at https://sustainabledevelopment.un.org/content/documents/612558-Inter-%20and%20Trans-disciplinary%20Research%20-%20A%20Critical%20Perspective.pdf [6 January 2019].